wer, Pollution, d Public Policy

Issues in
Electric Power Production,
Shoreline Recreation,
and Air and
Water Pollution
Facing New England
and the Nation

M.I.T.
SEA
GRANT
PROGRAM

MASSACHUSETTS INSTITUTE OF TECHNOLOGY

National Sea Grant Program

Sea Grant Project GH–88

Power, Pollution, and Public Policy

M.I.T. Report No. 24

Power, Pollution, and Public Policy

Issues in Electric Power Production,
Shoreline Recreation, and Air and Water
Pollution Facing New England and the Nation

Interdepartmental Student Project in Systems Engineering
at the Massachusetts Institute of Technology, Spring Term, 1970

Dennis W. Ducsik, Editor

The M.I.T. Press
Massachusetts Institute of Technology
Cambridge, Massachusetts, and London, England

Library of Congress Cataloging in Publication Data
Main entry under title:

Power, pollution, and public policy

([Massachusetts Institute of Technology. Sea Grant
Project Office] Report no. MITSG 71-8) (M.I.T.
report no. 24)
        "Index no. 71-108-Cid."
        Outgrowth of Project NECAP (New England Coastal Area
        Planning), a study conducted at the Massachusetts
        Institute of Technology in the spring of 1970.

        Includes bibliographies.

        1. Environmental policy--New England. 2. Electric
power production. 3. Pollution--New England.

I. Ducsik, Dennis W., 1946- ed. II. Massachusetts
Institute of Technology. III. Series. IV. Series:
Massachusetts Institute of Technology.

M.I.T. report no. 24.
HC107. A113E55   301.3'1'0974   71-173861
ISBN 0-262-04036-0 (paperback)

Second Printing, June 1972

FOREWORD

This volume resumes the practice of making available in book form the results of the effort initiated as a student design project under the subject "Special Studies in Systems Engineering." The last volume resulting from this subject grew out of an examination during the Spring of 1968 of the problems facing Boston's air and seaports and was published as Project BOSPORUS. During the Spring of 1969, the group explored the economics of alternative methods of transporting oil to markets from the newly discovered fields on the North Slope of Alaska. A significant thesis grew out of that work but it was decided not to publish a formal book on the study.

A different project was again selected for study during the Spring of 1970 and the work was carried through the oral presentation stage in May of that year. Since that time, Mr. Dennis Ducsik, who was one of the students in the subject, has conducted considerably more research on the topics considered during the term. He revised and expanded the original material and wrote the final drafts of Chapters 1 through 5 presented herein.

Each section of this document has been reviewed by at least one of the faculty members who participated in directing the study, but no attempt was made by the faculty to do more than offer suggestions. Thus, credit for the ideas as well as the form of presentation goes to the students and in particular to the editor, Dennis Ducsik.

William W. Seifert
Professor of Electrical Engineering
Professor of Engineering
in Civil Engineering
Massachusetts Institute of Technology

Cambridge, Massachusetts
June, 1971

PUBLISHER'S NOTE

The aim of this format is to close the time gap between the preparation of
certain works and their publication in book form. A large number of significant though specialized manuscripts make the transition to formal publication either after a considerable delay or not at all. The time and expense
of detailed text editing and composition in print may act to prevent publication or so to delay it that currency of content is affected.

The text of this book has been photographed directly from the author's
typescript. It is edited to a satisfactory level of completeness and compre-
hensibility though not necessarily to the standard of consistency of minor
editorial detail present in typeset books issued under our imprint.
The MIT Press

TABLE OF CONTENTS

ix

The concepts and analyses presented in this book were initially formulated during Project NECAP (New England Coastal Area Planning), a study conducted at the Massachusetts Institute of Technology in the Spring of 1970. The project comprised a semester's work in a single graduate subject by a multidisciplinary group of students from M.I.T., Boston University, and Wellesley College. The group consisted of thirteen graduate and five undergraduate students with backgrounds in law, economics, and six engineering disciplines, as shown in Table P.1. We felt that such an interdisciplinary framework provided an ideal forum for

Students	School	Department	Year
Ilyas Bayar	M.I.T.	Economics	Junior
Philip Byer	M.I.T.	Electr. Eng.	Senior
Larry Donovan	M.I.T.	Naval Arch.	Grad. Stud.
Dennis Ducsik	M.I.T.	Electr. Eng. & Management	Grad. Stud.
Robert Field, Jr.	B.U.	Law	Third Year
Sung Ling Ho*	M.I.T.	Nuclear Eng.	Grad. Stud.
Robert Jerard	M.I.T.	Mechanical Eng.	Grad. Stud.
Robert Jones	M.I.T.	Mathematics	Senior
Sandra Lynch	B.U.	Law	Second Year
Paul Mertens*	M.I.T.	Nuclear Eng.	Grad. Stud.
Steve Milligan	M.I.T.	Aero. Eng.	Senior
Richard Morse, Jr.	B.U.	Law	Third Year
Tom Najarian	M.I.T.	Mechanical Eng.	Grad. Stud.
George Neill*	M.I.T.	Nuclear Eng.	Grad. Stud.
Gary Petaja	M.I.T.	Mechanical Eng.	Grad. Stud.
Brunn Roysden*	M.I.T.	Nuclear Eng.	Grad. Stud.
Robyn Seitz	Wellesley	Economics	Senior
Robert Wolfe	B.U.	Law	Third Year

*These students participated in conjunction with Course 22.26 in the Department of Nuclear Engineering.

Table P.1 Participating Students--Special Studies in Systems Engineering

the discussion and examination of complex problems, problems that cannot be solved effectively without consideration of all their pertinent aspects and component parts. These, like the problems themselves, go beyond the boundaries of any single discipline and can only be confronted successfully from a multidisciplinary point of view, encompassing considerations of the social, political, economic, and technological issues that are necessarily involved.

It was proposed originally that the class take a long-range look at the role that the land-sea interface should play in the further development of the New England Coastal Area. The tentative goal was to lay out a regional development plan for shoreline utilization for the next 50 years, while taking a broad look at the problems of the area and the potential for solutions. During the first few weeks, however, the orientation of the study underwent a gradual evolution based on the attitudes and interests of the students involved. We decided that, within the allotted time, it was beyond the capability of the group to develop any comprehensive, detailed "master plan" for regional development. More important, we felt the need for a much different approach to planning: the utilization of a flexible and dynamic methodology to be applied continuously to meet the needs of changing times, rather than formulation of a long-range "master plan" that might be outdated before it could ever be initiated. With this orientation we then set out to tackle what we felt were "critical problem areas" facing society in New England and in the nation as a whole.

In the meantime, presentations by guest lecturers from government, industry, and universities helped to identify and describe the fundamental issues pertinent to our study. The students then formed subgroups to examine more closely particular focal points within the critical problem areas.

Although the orientation, management, and working structure of the course were left in student hands, the participating faculty, as listed in Table P.2, played a vital role as well-informed consultants through counseling, questioning, criticizing, and other forms of subtle guidance. Beyond this, however, there was no faculty veto nor direct control over the final product, responsibility for which rests with the students alone.

Faculty Member	Department
John W. Devanney, III	M.I.T.: Ocean Engineering
Michael J. Driscoll	M.I.T.: Nuclear Engineering
Tamar Frankel (Mrs.)	Boston University: School of Law
William Ryckman	Boston University: School of Law
William W. Seifert	M.I.T.: Civil and Electrical Engineering
David G. Wilson	M.I.T.: Mechanical Engineering

Table P.2   Participating Faculty

On the last day of the term, the students made a formal presentation of the results of Project NECAP to an invited audience
which included representatives from the business, government,
and academic communities who were actively concerned with the
problem areas under investigation. These results were then for-
mally written up in the form of term papers, which provided a
basis for the substantial amounts of follow-up research that went
into the articles presented herein. This subsequent effort as
well as publication of the manuscript was supported jointly by
the Henry L. and Grace Doherty Charitable Foundation, Inc., and
the National Sea Grant Program, Project GH-88, 1970-1971 project
element "Interdisciplinary Systems Design Course."

We are indebted to the entire participating faculty for their
enthusiastic support in this undertaking, and are particularly
grateful to Professor William W. Seifert for his active participation and able guidance throughout the entire effort. Also,
special thanks is due to Miss Virginia Root and Mrs. Louis Fischer
for their patience and perseverance in typing the large amounts
of material that went into the preparation of these articles.
Finally, we are grateful to Mr. Art Giodonni of the Electronic
Systems Laboratory Drafting Department for his meticulous effort
in preparing the illustrations found herein.

Dennis W. Ducsik

Cambridge, Massachusetts
June, 1971

INTRODUCTION

In the preparation of the papers contained in this book, the authors were guided by a general philosophy which maintains that the complex problems of modern society can best be attacked by taking, first, a *comprehensive* orientation and, second, a *social* orientation to the planning process. Both of these terms require some elaboration.

In speaking of a *comprehensive* orientation, we mean that we have approached in a systematic way, the identification of the relevant interactions among a *wide range* of technical, economic, social, and political issues that underlie that group of problems on which we focused our attention. Too often it seems that planners--be they economists, engineers, lawyers, or politicians-- operate within a relative vacuum, confining their efforts within the boundaries of their own discipline. As a result we note that the "best" approach frequently depends upon to whom one talks-- engineers tend to look to technological innovation, economists to fiscal policy, and politicians to legislative action. And too often, as debate goes on as to what constitutes the "best" approach, the problems go unsolved. It is hoped that the analyses presented in this book do not fall prey to these shortcomings. We have tried, within the limits of our resources, to develop a complete picture of the contexts within which certain problems arise and in which solutions must be carried out. We have tried to examine carefully the relevant aspects of each situation. We feel that this orientation is a most important component of any planning process that hopes to provide effective solutions to the complex problems that have become a part of modern American society.

In taking a *social* orientation we express our belief that one of the fundamental goals of planning should be the maximization of the "good life" as defined by an appropriate aggregation of the individuals and groups with whom the planner is concerned. Perhaps the most important question to be dealt with from this social perspective is: *are the results of the allocative mechanisms in our economic and political institutions consistent with the objective of improving the quality of life in American society, and are these institutions responsive to the needs and demands of our people?* We seek to discover if these mechanisms can be modified to achieve a more desirable overall balance in the allocation of our scarce resources, especially the most basic ones of *air*, *water*, and *land* whose protection has too often been of low priority in our preoccupation with affluence and growth. This suggests what constitutes the starting point of our methodological approach--the development of a framework within which the problems can be viewed and questions such as these can be answered.

The analytical framework that is developed in Chapter 1 is based on the mechanisms of the *private market*, the fundamental institution that we rely on for the allocation of resources in our free enterprise system. By examining these mechanisms, we can understand the root causes of particular problems as well as determine how the system might be altered to effect desirable solutions. In doing so, we develop basic *guidelines* that should be applied to decision-making in the public sector, pointing out important issues that must be faced when one tries to make decisions in the absence of the traditional institutional disciplines that have been relied upon so heavily in the past.

Within this framework, our methodology is simple. From a social standpoint, we compare the present situation in certain problem areas with one that seems more desirable to society, attempting to discover why the situation has come about. Then, with a comprehensive orientation, we examine the pertinent aspects of each problem, identify the critical forces at work, and focus on an area that deserves immediate attention. Our goal is to make some substantive contributions to the problem-solving efforts of those who deal with these problems in government, in business, in academic communities, and in private life. In addition, we have considered both the short- and long-term implications of the decision-making activities in each area of interest.

While a myriad of critical social, economic, and political problems face both New England and the nation today, we have chosen to focus on a certain few that have very direct effects on the quality of life for the great majority of American citizens. In particular, we have elected to concentrate on *environmental* issues regarding the misuse of our most fundamental scarce resources--the air, water, and land that comprise our natural environment. In Chapter 2, we focus on the critical area of electric power production and its associated difficulties in the areas of environmental degradation, land-use conflicts, and construction delays. In Chapter 3, we examine the crisis in shoreline recreation as a fundamental problem in the allocation of coastal land by the private market and localized political decisions. In Chapter 4, we probe the complexities of the air pollution problem with particular emphasis on the question of sulfur oxide emissions and their effective control. In Chapter 5, we dicuss the problem of pollution in Boston Harbor, paying particular attention to municipal responsibility in improving the water quality so that the full potential for useful harbor development can be realized. Finally, in Chapter 6, we take a futuristic look at what kinds of political reorganization might be required to effectively manage, in the long run, problems such as air and water pollution and land use that are inherently *regional* in nature.

In all but the last chapter, we have dealt primarily with short-term issues of immediate importance. However, we are also careful to note that, while short-term solutions can still be found, we must realize that these are usually stopgap measures at best and that the fundamental causes of all these problems are rooted in the growing size and wealth of our population. While the problems of environmental degradation can be traced to imperfections in our present allocative system, corrective action has been set aside for too long in favor of the American love affair with more, bigger, faster, etc.--adjectives that are not necessarily associated with *better*. We have reached the point where increases in quantity can no longer be considered the equivalent

of increases in quality. Unless some basic attitudes in the
"American Dream" are altered, we will inevitably face hard-to-
resolve tradeoffs such as between breathing clean air and having
enough power to satisfy increasing demands. The name of the
game is *saturation*, and time is growing short.

On this note we move into the body of our work. While we
believe that the only real solution to most of our growth-related
problems is the stabilization of population, we recognize that
other techniques must be employed to meet the crises that exist
right now. Hence, our goal in presenting these articles is to
suggest the means by which we might alleviate some of the most
severe problems we face today. At the same time, we hope that
the complexity and severity of the social problems here discussed
will alert all those who involve themselves with this book to
the dangers that lie ahead along the path of unbridled growth.

CHAPTER 1

THE FRAMEWORK FOR ANALYSIS

by

Dennis W. Ducsik

ABSTRACT

An analytical framework is a useful policy-making tool by which complex environmental problems can be defined, their causes identified, and alternative solutions evaluated. The misuse of valuable environmental resources such as land, air, and water can best be understood within the context of the institutions we rely on for the allocation of scarce resources. These consist of the economic environment of the *private marketplace* operating under the constraints imposed by the *political arena*.

In a properly-functioning market, the price-profit mechanism will bring about an efficient allocation of goods and services consistent with the values of society, as expressed by a willingness to pay. However, if certain conditions are violated, markets fail to appropriate resources in a manner that provides maximum benefit for society. This gives rise to the need for collective action. The problems of the environment are direct results of the failure of the private market to efficiently allocate our land, air and water resources.

In addition to market imperfections, there are political forces that inhibit the resolution of environmental conflicts. While these problems often spill over from one political jurisdiction to the next, there is no corresponding flow of governmental authority. Another difficulty is that decisions are made at localized levels where parochial considerations usually take precedence over the interests of broader-based constituencies.

This framework suggests some guidelines for decision-making in the public sector. The first issue to be confronted by policy-makers is determining the proper sphere of action in which a problem should be handled. To do this, it is important to realize when the private market will or will not work well. Beyond this, it is important to comprehend the qualitative functional difference between the public and private sectors in areas such as the determination of the public interest. It is not enough to simply reject market allocations--we must be confident that the new allocative mechanisms (collective action) will be better than the old.

5

CHAPTER 1

THE FRAMEWORK FOR ANALYSIS

I.   INTRODUCTION

The purpose of this chapter is to provide an *analytical frame-work* within which important environmental problems can be de-fined, their causes identified, and alternative solutions evalu-ated. Such a framework is always a useful tool in the making of public policy since it gives the decision-maker a convenient reference through which he can grasp the causes, interacting elements, and effects of proposed courses of action associated with complex social problems. In this book, we have chosen to view some of the high-priority issues facing New England and the nation from the standpoint of *social balance* and *efficiency* in the allocation of scarce resources. These are the fundamental concepts upon which our framework for analysis is based.

An *efficient* allocation of scarce resources can be defined as one that is most consistent with the aggregated goals and values of the whole of American society, as expressed by a willingness to pay for goods and services in a private market economy. This means that resources should be allotted to the production of goods and services in proportions that are determined by how much of each good society demands. This formulation of the concept of efficiency incorporates the notion of social balance, i.e., *the relative proportions of goods and services that are produced in our economic system must accurately reflect the values that society attaches to each good.* This implies that the allocation of scarce resources also must conform to the relative desires and interests of the general populace. Hence, our fundamental objective of improving the overall quality of life in America can be considered synonymous with achieving the goal of efficiency with social balance in the allocation of scarce resources. This will be discussed in more detail in the following section.

We have taken this orientation because environmental difficulties in the areas of electric power production, shoreline recreation, and air and water pollution are a direct result of a
misappropriation of our most basic natural assets--unspoiled land,
fresh air, and clean water--in the absence of any effective articulation of their value to the American people. The lack of responsiveness to such value on the part of our allocative system
has led to the widespread misuse of these scarce land, air, and
water resources. Recognition of this fact leads us to an examination of the allocative mechanisms that we rely on to articulate
and respond to our overall goals, values, and interests.

In this country, the allocation of scarce resources has always been determined within the economic environment of the *private marketplace* operating under constraints imposed by the *political arena*. Historically, the early concept of laissez-faire
and an unregulated market has been modified to the point that
today it is generally acknowledged that there exist *three* broad
areas within the economic environment in which goals of public
policy should be defined and collective action taken in the public interest. These areas include:

1)   economic stabilization and growth
2)   the distribution of income
3)   allocative efficiency and social balance

In the first two categories, governments have customarily exercised their influence by designing policies to combat inflation,
decrease unemployment, or transfer income to the underprivileged
through various social security programs. In this book, we are
concerned primarily with the third category. In seeking efficiency,
we need to develop a rationale for collective actions and an understanding of the consequences and implications of such actions.
We do this in the following section by exploring in greater detail the fundamental precepts of economic efficiency and its relation to social balance. This then will provide us with the framework within which we can formulate some general guidelines for
decision-making in the public sector and from which we can

approach the critical problem areas of particular interest in
the following chapters.

II. RESOURCE ALLOCATION BY THE PRIVATE MARKET

1. The Concept of Economic Efficiency

In speaking of an economic "good," we are referring to anything that society desires, be it physical, psychological, esthetic, or otherwise. Clean air and public beaches can be thought
of as goods in this sense, along with automobiles, television
sets, electric power, haircuts and other familiar commodities.
Since a limited amount of resources is available to our society,
each good can only be produced up to a certain maximum level,
assuming the levels of all other goods are held constant. We
will achieve this level only if we make efficient use of all the
resources at our disposal, i.e., labor, technology, and natural
resources. If two or more goods are to be produced, many combinations of the levels of each good are possible with the efficient
application of resources--but *efficiency requires that to have
more of one good implies that we must have less of others.*

This concept is illustrated in Figure 1.1, a graphical representation of what is known as a *production-possibility* curve for
a hypothetical economy in which only two goods are available to
society--electric power and coastal land use for recreation. The
curve shows that if no coastal land is devoted to recreation
(Point 1), we can obtain a certain very high level of power production by locating plants at the coast (where the required
cooling water supplies are available). Similarly, if no power is
generated, all the coastal land could be used for recreation
(Point 2). Between these two extremes there exist many production
combinations of the two goods (Points 3,4,5, etc.) all of which
represent an efficient use of the land, labor, and technical resources available.

This points to an important concept, i.e., there is no single
most efficient production combination with its corresponding allocation of resources; rather, a distinction is made between *ef*-

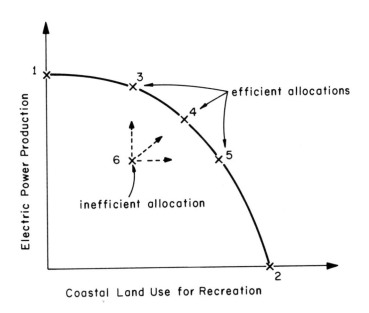

Figure 1.1 Production Possibility Curve for a Two-Good Economy

$\{ieient$ and $ine\{\{ieient$ allocations. An inefficient allocation
of resources implies that we could have more of one good $without$
reducing the amount that we can have of the other good (assuming
that society always prefers more of a particular good to less).
Point 6 in Figure 1.1 represents an inefficient allocation since
it does not lie on the production possibility curve; hence,
society could move toward point 3, 4, or 5 and be better off.
This means that a more efficient appropriation of resources could
enable us to have more power without decreasing the amount of
recreational land available, or vice versa. This might come
about, for example, if previously unused technological capabilities were employed to increase the generating capacity of a given
station; if new construction methods were developed so that a
power plant required less coastal acreage; or if imperfections in
the market were corrected to avoid inefficient use of coastal land.

In reality, the production-possibility curve is a multi-
dimensional surface, a complex representation of the production
combinations of all conceivable goods and services. However,

the concepts of efficiency and inefficiency and their distinction
remain unchanged. It remains to be seen how the values of society
(social balance) can be included in the analysis. The question
is: which of the efficient production points is most consistent
with the overall desires of the American citizenry? The answer
to this question lies in the theory of the properly-functioning
market.

2.   The Properly-Functioning Market[1] and Social Balance

The private marketplace is the mechanism through which society exercises the choice among the combinations of goods it might
have. Taking the distribution of income as given, a *properly-
functioning* market translates aggregated personal values into desired amounts of production through the workings of the price-
profit system. The *price* mechanism brings about effective proportional representation of individual values--as reflected in a
willingness to pay--through the "vote" of the dollar. The
profit-incentive mechanism brings about efficiency through the
flexibility of decentralized, unregulated decision-making which
allows resources to be channeled to their most valued use. In a
properly-functioning market, competition among buyers assumes that
goods and services will be allocated in conformity with the relative desires and abilities of the participants to pay. *If certain basic conditions are met*, there will exist a set of market
prices such that profit-maximizing firms and benefit-maximizing
consumers who respond to those prices will automatically direct
the economic system into an efficient allocative position that is
most consistent with the aggregated values expressed by society.
We term this position the optimal allocation since it is both
efficient and consistent with society's desires. It is important
to note that there is an optimal allocation for every distribution
of income since values will change in general as the income distribution is varied.

If the private marketplace always functioned in the proper
way (all assumptions and conditions fulfilled), the necessity for
collective action to help bring about efficiency would be mini-

mal. There are those who argue[2] that the unregulated market system in almost all instances benefits the public in attaining
efficiency since the market can provide "unanimity without conformity" and "effective proportional representation" of aggregated
individual preferences. However, even the most loyal defenders
of the competitive market admit that there are certain circumstances
in which markets fail to provide certain desired outputs and
underproduce others. Such inefficient situations come about
when the conditions and assumptions upon which conclusions are
made about the effectiveness of a market system are *not* satisfied
in reality! It is important now to determine the circumstances
in which the private market does not work well. We seek to
discover what steps might be taken within the institutional
environment to correct for the inefficient production of goods
(due to a misallocation of available resources) and bring about
efficiency with social balance.

3. Market Failure

What then are the conditions necessary for markets to function smoothly and which, when violated, lead to market failure?
The ones most germane to this analysis are as follows:

1)   information must be available
2)   the price of a good must reflect the total social
      cost of its production
3)   the characteristics of goods and services must
      meet certain criteria

1)   *Information* is an important factor in any efficient
operation. Producers need knowledge of available technologies,
demand, potential markets, and the costs of inputs. Consumers
need to know what goods are available and what their characteristics are. Both need to know the relevant set of prices.
In some instances, information may be scarce, costly, unreliable,
or hard to understand, interpret, and evaluate without special
training.

2)   The *price* of a good must reflect the true cost of lost

opportunity to society, i.e., the value for other uses that is
given up when the good is consumed for one particular use. The
efficiency-seeking mechanisms in a properly-functioning market
require that the social benefits of consuming a particular good
must exceed the social cost of lost opportunity (thereby making
society better off). It is this total cost to society that must
be reflected in the price of the good.

3)   The *characteristics* of goods and services must meet
certain criteria in order to be suitable for allocation by a
private market system. One such criterion is that there be
no violation of the *exclusion* principle: i.e., pricing demands
the possibility of cost-free exclusion of non-buyers from the
use of the product. This may be technically impossible or expensive. The classic example is national defense, where use by
one person neither diminishes nor excludes availability to others.

We can generally classify goods that violate the above conditions as *public* or *collective* goods since they are in need of
some institutional involvement to correct allocative deficiencies
of the private market. Public goods are most often characterized
in the following ways:

1)   It is impossible to price the good correctly due to
difficulties in measuring the amount of benefit derived and in
translating this into revenues--*true social cost not reflected in
price;*

2)   The basic values of society make it desirable to keep
the good out of the private market system. Police and fire protection and public education are examples of public goods that
could be produced in the private market but are not, since society places large value on the idea that everyone should derive
equal benefit from such institutions regardless of income distri-
bution--*no exclusion should exist;*

3)   There are *externalities* or *side effects* associated with
the production and/or consumption of the good. These effects
come about when the production of certain goods affects other decision-making units which are *not* doing the producing or the con-

suming. Side effects are not included in the price of the good
since there is no mechanism by which the external costs to society can be returned to the producer as the cost of a factor input
to production--*pollution is the classic example.*

These characteristics point to the breakdown of the price
mechanism in the allocation of public goods since they all involve violations of the conditions of a properly-functioning
market. The crucial point that must be reemphasized is that
frequently the total opportunity costs to society are *not* reflected
in the price of goods. Although the overall social costs of
having an individual consume/produce (or not consume/produce)
a particular commodity may exceed his private costs, he will
base decisions only on his private costs. The private market,
left alone, tends to produce too many *private* goods and too
few *public* goods. This happens because the public goods are
undervalued by the private market and are unable to compete
on an equal footing with other goods in the allocation of scarce
resources. For this reason, some form of collective action
is required in order to maintain social balance and achieve
an efficient resource allocation consistent with the overall
goals and values of society!

We are now in the position to make the connection between
the scarce natural assets of unspoiled land, fresh air, and clean
water and their allocation in a private market economy. It will
soon become clear that these assets should be considered to be
public goods in every sense of the word.

III. ENVIRONMENTAL RESOURCES AND MARKET FAILURE

The rationale for regarding our scarce environmental resources of land, air, and water as *public goods* is based on the
fact that there are substantial undesirable *social side effects*
that accompany their use in our economic system. Since these
effects are in no way evaluated and translated as a factor input
to the production process, the private market has failed to allocate our natural assets according to the true values of society

as a whole. Market failure in the critical problem areas that
we have chosen for study has come about as a result of such
externalities in the allocation of coastal land, urban air, and
ocean water. These resources can no longer be considered "free"
and limitless in a rapidly-expanding economy since their unregulated usurpation has led to significant, unaccounted-for disbenefits to the health and well-being of a great number of American
citizens. Each critical problem area that we have studied can
be shown to qualify as a matter for public concern and collective
action due to market failure in the presence of externalities.

1. Coastal Land Use for Public Recreation

Historically, those uses that could pay the highest prices
for coastal land have preempted most of the shoreline. These
uses have most frequently been for industrial and commercial development, housing, and private recreation, all of which have for
a long time been well established in the competitive marketplace.
The allocative mechanisms of the market have functioned well with
regard to the distribution of coastal land among these competitors. Unfortunately, public recreation has never been able to
participate effectively in the competitive process since the bids
for land from other uses have far outstripped those for public recreation. The result is that only a small percentage of
the entire American shoreline--about five percent--is in public
hands for recreation. This has come about because there has been
no effective mechanism by which the recreational, esthetic and
ecological values of shoreline resources to an entire region can
be reflected in the price of coastal land. The greatest difficulty in this regard has always been to put a price on certain
values, much less find a way to translate these values into revenue. Yet the private market demands that these be done by any
use which seeks to compete for control of coastal land. Our
state and local governments, subject to increasing financial
stress and the pressures for continued economic development,
have been unable to adequately represent in the economic arena
the true value of shoreline recreational opportunities to their

constituents. Thus the cost of lost opportunity for recreation
to regional society is *not* effectively represented in the competitive bidding for coastal land. This *side effect* of lost opportunity indicates that our shoreline resources must be allocated
by some mechanism other than that provided by the private market
in its present form.

2. Air and Water Pollution

The problems of air and water pollution are classic examples
of external effects and occur largely because of the difficulty
in imposing direct monetary responsibility on those who benefit
from pollution. The desire at every private level to minimize
costs, coupled with the traditional notions regarding air and
water as limitless resources to be used freely by all who desired
to do so, has led to a gross misuse of these environmental assets.
This misuse has given rise to the all-too-familiar harmful side
effects that pollution has on human health, plant and animal
vitality, and the overall human environment. In addition to
these immediate disbenefits, pollution can destroy some of the
major productive assets upon which our future prosperity rests.
These effects most often accrue to persons other than those
who are directly involved in the production or consumption process
of a particular good. Yet there exist no mechanisms by which
the costs to society associated with these effects can be returned
to the producers as a factor input to production.

A good example of the externalities associated with pollution
is the case of a paper mill located in a town on a rural river.
Assuming the river is not privately owned, the owners of the
mill will consider the local air and water supplies to be free
receptacles for the discharge of their effluent wastes. Sludge,
chemicals, and heated water may be emitted into the river, while
smokestacks may spew out gases containing pollutants(such as
sulfur oxide and particulate matter) as well as creating an
offensive odor. While the air and water may be free to the
paper mill, they are certainly not so to the residents of the
town or of a community farther down the river. These people

pay for the air in the form of decreased health, greater maintenance
bills, and an overall degradation of their physical environment;
and they pay for the water in the form of decreased recreational
and esthetic enjoyment since the discharges of the plant may foul
beaches, destroy fish and other wildlife, or create unsightly
slicks. Yet again there has been no way for these costs to society
to be transferred to the paper mill as a true cost in their manufacturing process. If this could be done, the paper mill would
look for an alternative disposal scheme that would be consistent
with the values of the water to the surrounding area (assuming
that the social cost of the pollution is greater than the cost of
abatement).

3.   Electric Power Production

An important factor in the recent difficulties faced by
electric power companies in meeting the rapidly-increasing demands of urban residents has been the unwillingness of the general public to accept any longer the harmful side effects that
power production has on our environmental resources. Fossil-
fired generating stations are heavy contributors to the problems
of urban air pollution; nuclear plants create serious temperature
increases in cooling waters which endanger the ecological systems
of rivers and bays; and the location of power plants at coastal
sites excludes large portions of the intrinsically valuable
shoreline from use by the public for recreation, esthetic enjoyment, and wildlife preservation. Thus, the question of electric
power production has associated with it all the external effects
on environmental resources and their use by the public that we
have discussed in the previous sections.

4.   Summary

The problems that we have examined in this book, including
shoreline recreation, air and water pollution, and electric power
production, all fit within the economic framework of market imperfections in the form of externalities that lead to unaccounted-
for social costs. Since the costs to society associated with
these side effects have *not* been articulated in the private market,

serious misallocations have occurred that point to the need for
collective action in the public sector. This then leads us to a
discussion of the political environment within which the private
market operates. We will find that in this arena also there are
barriers to effective action. By discovering the political as
well as economic shortcomings in our allocative system, we will
set the stage for the development of *guidelines* for decision-
making in the public sector with regard to the careful allocation
of our scarce environmental resources.

IV. POLITICAL BARRIERS TO EFFECTIVE ACTION

The preceding sections have considered environmental problems from a primarily economic point of view. In addition to
the difficulties posed within this framework, there exist some
serious political barriers to effective solutions through collective action. Such obstacles are the result of two interacting
forces:

1)   Environmental problems such as coastal land use and
air and water pollution are *not* restricted to town lines, state
borders, or other political boundaries--they *spill over* from one
political jurisdiction to another;

2)   Political decisions controlling the allocation of environmental resources that may affect an entire region are often
made by *local* governmental bodies, who weigh costs and benefits
as they apply to the local community only.

Thus the problems of environmental resource allocation exhibit a
common political nemesis--"the stifling effect of jurisdictional
boundaries which, by a curious osmosis, permits the diffusion of
problems throughout the region, while blocking any corresponding
flow of governmental responsibility."[3]

1. Jurisdictional Spillovers

Spillovers between neighboring political jurisdictions are
commonplace. The effluent discharges from chemical plants and
oil refineries in northern New Jersey contribute substantially

to the foul air over New York City. Discharges of heated water
and industrial wastes from power and manufacturing plants located
on the Connecticut River in New Hampshire affect the fish life,
recreational activities, and other uses of the water by neighboring Vermont. Similar situations can be found whenever two or
more jurisdictions are situated on a common body of water, are
affected by the same air masses, or share in the use of a unique
recreational land resource. In these situations, when problems
arise that adversely affect the parties that share in an environmental resource, there is a common benefit to be derived in
taking collective action on a cooperative basis to ensure an allocation consistent with the interests of all who are involved.
However, when benefits accrue primarily in one locale and costs
in another, *there is often no political framework within which
the resulting conflict can be resolved*. Pollution laws enacted
in one municipality do not apply in another. Regional authorities seem to lose their "teeth" when real conflicts of interest
arise between states. And even when one political unit decides
to take action, its efforts may be negated by a lack of cooperation from adjacent governmental bodies. For example, for a long
time New York City was unable, without resorting to legal means,
to obtain compliance with its pollution regulations from New
Jersey polluters, even though an estimated one-third of New York
air pollution originates in New Jersey.[4] In another case,[5] a
federally-sponsored multistate conference dealing with the cleanup of Lake Erie was temporarily disbanded when the governors of
New York and Pennsylvania withdrew their representatives on the
grounds that the problem was a matter for the states to solve on
an individual basis. A final example[6] is that of the New England
Interstate Water Pollution Control Compact, a regional authority
authorized in 1947 to classify sections of rivers and streams
according to a scale of potential water uses. Recognizing the
various problems presented by water pollution, this regional body
analyzed each situation on a cost-benefit basis and adopted a formal
classification system. After approval of classification, each
state was to hold the responsibility for obtaining action by munici-

palities and industries for the installation of water treatment
works. Although the number of these treatment plants was reported to have increased, by 1965 the quality of the rivers and
streams in New England had not improved substantially. This happened because the regional body had no power beyond the recommendation stage. Most treatment plants that were built provided only
primary treatment (which removes only a small percentage of many
harmful pollutants), and in cases where a technically complex
and expensive treatment process was required (often the case with
many dangerous industrial wastes), *no* action was taken. There
was no political means by which pressure could be brought to bear
on the states involved to require more effective abatement programs within their jurisdiction.

2. Localized Political Decision-Making

The problems of jurisdictional spillovers would be relatively nonexistent if localized decision-making units were to consider the costs and benefits that accrue to *all* who are affected
by a particular resource-consuming enterprise. But this does not
happen generally, as can be illustrated by looking at the decision-making process involving, perhaps, some coastal zone project
such as the location of a power plant on a beach shoreline. It
is important here to distinguish between two types of benefits
(or disbenefits) of such a project--*direct* and *indirect*. Direct
effects are those that accrue to the consumers affected by the
project--the user of the power supplied, the former bathers on
a closed beach, the breathers of polluted air, the viewers of
marsh wildlife, etc. All of these effects are felt both by the
local community *and* by the regional society. Yet only those
benefits (or disbenefits) that accrue to the *local* populace enter
into the decision. The community may be willing to give up beach
or bluff property to have a power plant, but this may not be an
efficient allocation for that resource on a regional basis. However, the "votes" of the region are not counted--only those of
the local community affect the decision!

We might ask why a community would be willing to give up

this valuable property in such a way. The answer is that the
local community within its own particular economic and political
context is also subject to a second type of benefits--*indirect*
or *secondary* effects. These effects accrue to the suppliers of
the resource that make the investment possible. Construction
workers who build the plant will spend a substantial portion of
their paychecks in the locale of the plants, certainly benefiting
local merchants, doctors, and bar owners. These people, in
turn, spend some of this money in the locale, and so on, in
the traditional multiplier effect. Values that arise in this
manner are also called *parochial benefits* and include the net
effects on job availability, local payrolls, retail earnings,
and the broadening of the tax base (usually a very powerful
factor). For the local community, these benefits are very real;
but considering the regional economy as a whole, they are not
net benefits since parochial effects associated with one location
will be about the same as those associated with an alternative
site (barring large unemployment differentials). Thus, parochial
benefits represent a transfer payment from one place in the
economy to another, with no net regional benefit associated
with the choice of site (even though there *is* a net benefit
to the particular community chosen). Yet parochial benefits
can be overwhelmingly important to political bodies representing
the local community. As a result, a local community can rationally
view a project in a very different manner from that in which
the regional economy as a whole views it. The region and the
local community feel positive and negative direct effects--the
community alone feels the positive parochial effects. These
added benefits may persuade a community to act in its perceived
self-interest and approve a power plant siting, with *no* consideration of the negative direct effect to the region as a whole, i.e.,
the loss of a valuable stretch of beach for recreational use.

3. Concluding Remarks

Within the public sector there are certainly many other
political barriers to solving environmental allocative problems,

including bureaucratic inefficiency and sluggishness, conflicts over jurisdictional prerogatives, and susceptibility to various forms of narrow political pressures. However, the two difficulties presented here are felt to be at the core of the contributions made within the political arena to the misallocation of environmental resources. These two problems have been dealt with specifically in Chapters 3 and 6 of this book. Chapter 3 deals with the problem of localized political decision-making with regard to recreational resources in the coastal zone. Chapter 6 is addressed to the general problem of jurisdictional spillovers and develops a possible structure for a *regional* government in the New England area.

At this point, we have completed the development of an economic and political framework from which we can view the problems addressed in this book. Within this framework we can now formulate some general guidelines to assist in the decision-making process in the public sector.

V.   GUIDELINES FOR DECISION-MAKING: THE ROLE OF GOVERNMENT

We have shown how an unregulated private market within a political environment of localized decision-making tends to overproduce private goods and underproduce public goods--such as pollution abatement, public beaches, etc.--through a failure to efficiently allocate resources in the presence of externalities and unaccounted-for social costs. These costs accrue in the form of lost opportunity to persons *other* than those who are doing the producing or consuming (or who otherwise make allocative decisions). Hence, society may choose to *reject* market solutions to allocative problems involving public goods (such as land, air, and water) out of a concern for the proper valuation of side effects like damage to the quality of man's natural environment. This establishes a firm basis for collective concern and public action, but general guidelines by which governments should carry out this action are yet to be determined. In this section we attempt to develop such guidelines and comment on their applicability to the topics discussed in the remaining chapters of the book.

There are two fundamental questions that must be answered regarding decision-making in the public sector. The first of these is, what is the proper *sphere of action* in which a given problem should be handled? The second is, *how* should we go about making decisions within the appropriate sphere of action? Each of these questions has both economic and political aspects, as one might expect in view of the dual nature of the environmental problems themselves.

1.   The Sphere of Action

In speaking of a proper sphere of action, I mean that there is a particular *institutional environment* within which certain problems can be dealt with most effectively. In many cases, the environment of the private market and localized political policy-making is perfectly adequate for the making of allocative decisions. But when this environment is found to be deficient, as we claim it to be in the case of environmental resources, it must be modified or replaced, based upon a careful examination of the available alternatives, both economic and political. In the political arena we must look carefully at which level--federal, state, local--or which combination of levels is best suited to manage the problem, and whether or not some form of reorganization is needed. In the economic arena, the possible alternatives include: 1) reliance on some sort of *adjusted* market system; 2) pure collective action (economic or political) *outside* the private market; or 3) some *combination* of the two.

In choosing among these alternatives, there is considerable advantage in knowing when markets do and do not work well. In addition, it is of major importance to understand the *qualitative functional difference* between the public and private sectors. Government policy-makers must determine how (if at all) the market can be revised to do the job, and what actions should be taken if the market cannot be adjusted properly. Convincing arguments can be made to the effect that market adjustment is preferable to most other collective actions on the grounds that it preserves the clear advantages of *free* and *decentralized* decision-making, greater *flexibility* in attaining efficiency, and more effective *proportional representation* of individuals' values through the dollar "vote." If such an adjustment is not possible, policy-makers may attempt to *simulate* the market to determine what outcomes would result if the market were working under the proper conditions and then take steps (through legislation or public spending) to bring about these desired outcomes. If this fails, government may find it necessary to take pure collective action in the form of prohibitive laws or regulating agencies to directly control an otherwise unmanageable situation. The proper sphere of action as discussed here must be determined by a close examination of the nature of each particular environmental problem and the availability of appropriate public policy tools.

2. The Decision-Making Process

The second major consideration pertinent to the management of environmental resources in the public sector is the question of *how* decisions are to be made regarding allocation among competing uses. If we conclude that the mechanisms of the private market are to be abandoned, then the state and federal management authorities must have some alternative means for determining what is an efficient allocation of those resources. This must necessarily involve the determination and articulation of the *public interest*. In the private market, goods have a mechanism (price-profit system) whereby the demands of individuals can be felt; when the aggregate of individual demands is high enough, private producers will attempt to satisfy those demands. Thus, many individual preferences can be satisfied since each individual's "vote" (in dollars spent) goes relatively far in determining the available supply. Whenever enough individuals want something at a price, there is an incentive for someone to produce it at a profit. Public goods differ in that private markets fail to respond to the entire range of individual demands, giving rise to a need for collective action. The question is, how can individual preferences for these goods be summed to determine if the aggregate benefit is sufficient to justify the

total cost? This is a central question in the area of welfare
economics, and the resolution of the issues involved must ultimately play an important role at the federal and state levels in
the formulation of management policy concerning the nation's environmental assets.

A number of theories[7] have been set forth involving this crucial determination of the public interest. The point of view of
an *aggregated social welfare function* holds that society maintains a hierarchy of priorities based on collective values,
inviting a search for the articulation of these priorities within
the political process. A fundamental question to be dealt with
in this regard is: Are these social priorities effectively articulated through the democratic political process as it now exists
so that decision-makers are adequately equipped to act in the
public interest? Another point of view is that of *willingness to
pay*, which holds that the maximum amount of resources that consumers are willing to pay for a good is a valid measure of its
value. This can be expressed as a willingness to pay additional
taxes, user fees and other charges; to give up the consumption of
certain goods; or to pay a higher price for other goods. The
primary objection to this scheme is based on the difficulty in
measuring the willingness to pay for *public* goods that are not
"unitized" and whose benefits to an individual are hard to determine. Cost-benefit analysis uses willingness to pay and appears
to have, in some cases, the potential for effective simulation of
the working of a properly-functioning market in the allocation of
public resources, usually on a project-by-project basis.

Whatever the method used to determine the public interest,
an important issue that must be dealt with is that of *equity* in
the determination of who should benefit and who should pay, and
in what amounts. Should two plants that discharge the same
amount of air pollutants but affect the environment differently
(because they are in different locations) be treated equally?
How do we transfer the costs of water pollution from those who
are not involved in the production/consumption process to those
who benefit directly from the "free" means of waste disposal?

Should we charge user fees at a city beach which might help finance the project but have an effect on the ability of poor people
to use the facility? These are just a few of the questions of
this nature that policy-makers must deal with before giving approval to any particular proposal. It must also be realized that
government policies themselves may bring about side effects that
must be accounted for within the framework of overall goals.

While there seem to be no clear-cut indications that any
method of determining the public interest is superior to the
others, this is no excuse for inaction--attempts must be made to
determine the values of society. Perhaps the answer lies in some
combination of the viewpoints of *representative political consensus* (based on overall social priorities) and *cost-benefit analysis* (based on willingness to pay) as effective measures of the
public interest. The important point is that some determination
must be made, at all levels, before we can claim that any new
framework for environmental resource allocation is *better* than
the old one of the private market and localized political decision-making. The fact that confrontations with the difficulties
in articulating the public interest have been avoided in the past
has allowed the environmental problems to continue unchecked for
so many years.

Having now completed the description of a framework for
analysis and a general orientation from which to view the decision-making process in the public sector, the final objective is
to show how this framework has been applied to the critical problem areas.

3. Applicability to the Critical Problem Areas

The reader will find in the subsequent chapters that the
appropriate spheres of action for the chosen problem areas cover
a broad range of policy alternatives. In Chapter 2, we propose
that many of the environmental problems related to electric
power production can be solved through *technological innovation*
and can only be managed as part of a coordinated effort, at the
federal level, aimed at formulating a national energy policy.

In Chapter 3, we have concluded that the problems of shoreline
recreation and coastal land use can only be attacked by more
broadly-based governmental units than local communities, while
the *private market must be modified or else abandoned* in the allocation of valuable shoreline resources. In Chapter 4, we have
concluded that the control of sulfur oxide air pollution at the
local level can best be effected within the present political and
legal environment by *adjusting the constraints on the private
market* to correct for allocative deficiencies. In Chapter 5, we
have stated that the serious pollution of the water in Boston
Harbor could be significantly reduced *using proven and commonly-
used technology* for the disposal of sludge from waste-treatment
plants. Finally, in Chapter 6, we have directly addressed the
problem of jurisdictional spillovers by proposing the *reorganization of New England government at a regional level.*

   With regard to the determination of the public interest in
the projects undertaken, we recognize that the usefulness of
alternative techniques is a function of the overall magnitude
that a given problem has reached. Since the problems of environmental misuse have gone unattended for so long, they are at the
point where something must be done *immediately* to reverse the
trends of continuing degradation. In this situation, we feel
that the overall values of society can be clearly articulated
through political processes, especially in these times when
public concern for the environment is so effectively mobilized
and widely publicized. Citizen groups across the nation have
acted to halt the construction of power plants, to aid in cleaning
up rivers and oil-fouled beaches, and to protest the contamination
of the air we all breathe. Certainly, issues of environmental
quality are among those in the forefront of domestic priorities
in America today. Thus, we have not perceived the need to go
beyond this generalized articulation of the public interest
through representative political consensus. The very nature
and scope of the problem areas that we have chosen to examine
are such that we cannot hope to achieve the optimal allocation
of our environmental resources overnight; rather, we seek to

take major strides in the *right direction*, one that will bring
to a halt the dangerous misuse of our valuable natural assets.

Alternatively, an accurate determination of the public interest becomes crucial when the scope of a given problem is more
narrow. Decisions as to the location of a particular power plant
on an estuary or the building of an oil refinery near a coastal
beach demand that careful attention be given to the evaluation
of the values of people in the region affected. As progress is
made in the reduction of pollution, sophisticated techniques may
be necessary to determine at what level society is willing to
live with pollution and beyond which it is not willing to make
the sacrifices necessary to reduce it further. The same sort of
thing applies to electric power generation. There will always
be a point at which these kinds of tradeoffs will be necessary,
and we must be equipped with the policy tools that can effectively confront such issues when they occur. While we have not
felt the need to deal with these more detailed questions expli-
citly in our analyses due to the broad scope of the topics discussed, we recognize them to be among the most important public
policy-making issues to be faced in the management of the environment in the future.

4. Concluding Remarks

In concluding this chapter, I would like to emphasize one
important theme that is repeated throughout the book. Society
must come to realize that we are only cheating ourselves if we
equate the goal of a higher national output with the desire to
improve our overall standard of living. The public must become
aware that Gross National Product is only a measure of market
quantities at market prices and is not adjusted for the negative
influences of social costs due to side effects such as pollution.
It is time to set our goals according to some measure of our
real standard of living, in terms of *quality* rather than those of
quantity. We must recognize that the satisfaction of our increasing demands for goods and services often comes at the expense of the quality of our environment. We must somehow seek
out and accept a compromise with nature.

REFERENCES

1. For a more detailed discussion of market performance, see
    the articles presented in The Analysis and Evaluation of
    Public Expenditure: The PPB System, Volume I, Subcommittee
    on Economy in Government of the Joint Economic Committee,
    91st Congress of the United States, Washington, D.C., 1969.

2. Milton Friedman, "The Role of Government in a Free Society,"
    in M. I. Goldman, ed., Controlling Pollution, Prentice-Hall,
    Inc. (1967).

3. Harvey S. Perloff and Lowdon Wingo, Jr., "Urban Growth and the
    Planning of Outdoor Recreation," in Trends in American Living
    and Outdoor Recreation, U.S. Outdoor Recreation Resources
    Review Commission Study Report No. 22, Washington, D.C.
    (1962), p. 84.

4. Marshal I. Goldman and Robert Shoop, "What Pollution" (see
    Reference 2, p. 67).

5. Marshal I. Goldman, "Pollution: The Mess Around Us" (see
    Reference 2, p. 23).

6. Jared E. Hazelton, "Effluents and Affluents," New England
    Business Review, June 1965, pp. 2-9.

7. For a detailed discussion of the public interest, see P. O.
    Steiner, "The Public Sector and the Public Interest," in
    The Analysis and Evaluation of Public Expenditures: The
    PPB System (see Reference 1).

CHAPTER 2

OFFSHORE SITING OF ELECTRIC POWER PLANTS

by

Dennis W. Ducsik
Paul Mertens
George Neill

ABSTRACT

It is increasingly evident that utility companies are hard
pressed to satisfy the rapidly multiplying demands of modern
American society for electric power. While the technology associated with power generation is well-developed, the industry has
been confronted with major obstacles involving social, political,
and economic issues. The most severe controversies have been concerned with questions of environmental quality, public safety, and
land use priorities in relation to the selection and approval of
sites for new generating facilities.

In responding to this situation, we have investigated a new
concept: the siting of electric power stations at offshore locations. This innovation seems to hold great potential for the
effective resolution of land-use conflicts while having extremely
attractive features in the areas of environmental protection and
public safety.

Our analysis has been directed towards the short-term issue
of determining the technical, economic, and legal feasibility of
this concept. We have concluded that it is indeed feasible from
all these perspectives with no undue extrapolation of existing
capabilities.

What is needed at this point is a careful evaluation of the
offshore concept as part of a coordinated effort at the federal
level directed toward long-range planning and the formulation of
a national energy policy. At present, no centralized governmental authority is charged with the responsibility of considering
national priorities, both on the short- and long-term, regarding
this crucial issue of electric energy production. An examination
of the issues surrounding the implementation of the offshore concept would be an all-important first step that would draw attention to the increasing need for the formulation of a long-range,
national energy policy.

CHAPTER 2

OFFSHORE SITING OF ELECTRIC POWER PLANTS

I. INTRODUCTION

A number of factors contributed to the selection of electric power production as a focal point for investigation by the
members of this study group. Certain problems in New England
are shared with other regions of the country, such as the major
role played by power plants in water and air pollution and
the difficulties faced by the utility industry in satisfying
increased demands. These problems alone merit the degree of
attention devoted to this topic. In addition, the New England
coastal area suffers some specific disadvantages, making the
cost of power to both commercial and residential consumers in
this region among the highest in the entire country. This adds
additional aggravation to an already difficult situation,
making the question of power production an acute regional problem. On the other hand, it became clear early in the study
that the region also possesses certain assets, both inherent
and cultivated, that permit the development of a specific solution to this problem.

The proposed solution to the problem of electric power production in the New England coastal area consists of the use of
barge-mounted power stations, constructed in New England shipyards and towed to and moored at appropriate offshore sites.
The numerous interacting considerations that led to selection
of this concept and the unusually synergistic features of the
Barge Mounted Power Station (BMPS) are described in more detail
in the following sections of this chapter.

II. BACKGROUND

1. The National Situation

During the mid-1960's, the primary issues confronting producers of electric power were engineering and economic in nature,

based on the choice between the use of nuclear- or fossil-fueled
generating facilities. Proponents of nuclear power generation
emphasized the advantages of using nuclear facilities to reduce
the industry's major contribution to air pollution in dense
metropolitan areas, and to otherwise soften the environmental
impact that large generating plants have on their surroundings.
Another attractive feature was the promise of substantial savings
in the cost of fuel, a major component in overall power costs.
For one early (1965) nuclear plant, the estimated cost of nuclear fuel was 16.0 cents per million Btu,[1] as opposed to the
United States average cost of 25.2 cents for fossil fuel in
the same year.[2]

In 1965-1966, nuclear power made a sudden breakthrough into
the power production market; between January 1966 and October
1967, orders were placed for 40,000 megawatts of nuclear generating capacity. This surge was spurred by the extraordinarily
low prices of the Oyster Creek, Dresden, and Brown's Ferry
Stations in 1964-1965; by the prospect of low nuclear fuel costs
in the face of rising coal costs; by considerable increases in
the price of conventional firing equipment; by rising concern
for the problems of air pollution; and by the exciting technological potential for future use of atomic energy.

All these factors made nuclear power particularly attractive to New England in 1965, when 89 per cent of the power used
was being produced by fossil-fueled plants, at an oil cost of
34.4 cents per million Btu and a coal cost of 33.6 cents.[3]
Also, the serious air pollution problem in the Northeast Corridor provided added incentive for the growth of nuclear power in
the region. Hence, by 1969, nuclear power accounted for 8.5
per cent of the total generated power in New England, up from
2.4 per cent in 1965. Although this represents a sizable increase, nuclear power still holds a relatively small share of
the total power market.

At the end of 1967, the Atomic Energy Commission (AEC) engaged
Mr. Philip Sporn to prepare an analytical report on develop-

ments in the power industry from 1962 to 1967. The results of
his investigation were interesting; he reported that, at the end
of 1967, nuclear energy had not progressed appreciably since its
initial breakthrough two years earlier. "Thus, despite increases in unit sizes at nuclear plants of as much as forty per
cent, costs at the newest plants were not expected to be less
than the 1965 level of 22 to 24 cents per million Btu."[4]
Sporn attributed this to more realistic construction and fuel
costs than were quoted initially in the unusually intense competitive bidding among manufacturers for the Oyster Creek contract. The ambition to break into the nuclear market caused
many competitors to undertake "cut-rate" contracts, hoping that
losses associated with initial risks would be more than balanced
by future contract awards. Hence, by the time the TVA con-
tracted for its third nuclear station at Brown's Ferry in 1967,
costs had escalated considerably, soaring to 140-150 dollars per
kilowatt of capacity (as opposed to 115 dollars in 1966).
Nevertheless, at 22 to 24 cents per million Btu, "nuclear plants
are now regarded as being competitive in all parts of the United
States except those immediately adjacent to the coal-producing
areas. Even so, utility choices in favor of one or the other
fuel are, in present circumstances, by no means simple as there
are considerable uncertainties on both sides."[5] Hence, new
plant decisions are being subjected to increasingly stringent
analysis.

In addition to increased construction and fuel costs, other
factors have contributed to the loss of the competitive advantage enjoyed briefly by nuclear power in 1965-1966. One result
of the flood of orders for nuclear plants in 1965 was the "creation
of a number of manufacturing and industrial bottlenecks which have
led to big stretches in delivery time--six years is now quite
quite common and at least one case has been reported where a delivery time of eight years was quoted."[6] These delays involve
huge financial risks to the utilities, especially in these times
of continuing price escalation. Another factor has been the AEC's
stringent safety regulations and criteria that must be adhered to

in the design and construction of nuclear containment vessels, adding still more to the overall costs. Also, there has been increasing concern over the potentially harmful ecological effects of thermal pollution of water resources used for cooling purposes by nuclear plants.

Finally, there has been increasing concern of late regarding the issue of radioactivity.[7] Some scientists[8] have suggested that the AEC's minimum permissible radiation levels are too high. While this possibility seems remote in light of the extraordinary precautions taken by the AEC, it cannot be denied that such a statement has a sobering effect on some segments of the general public. A second and much more serious issue in this regard is the problem of how to safely dispose of the vast amounts of high-level radioactive wastes that will accompany the further development of nuclear power in this country. While the use of geologically stable salt beds seems to provide the best possible alternative at present, the long-term seriousness of the handling and disposal problem continues to make it a topic of great controversy at the national level.[9,10,11]

So we see that the initially bright prospects of nuclear power generation have been somewhat tarnished to the point that there is, at present, an impasse between it and conventional generation methods. The well-intentioned efforts of the power community to find alternate means of supplying power without damaging the environment have brought them full circle to face the same serious problems, only now they are doubly intense. In the meantime, we have found ourselves face to face with perhaps the most urgent crisis in the history of the electric power industry. For the first time, there is doubt that the industry's capacity will be able to keep up with the escalating per capita demands of our increasing population.

Some appreciation of the nationwide power production problem is a prerequisite to the more specific discussion of New England's regional problems. The rapidly increasing demands of our automated society for electric energy are relatively well-

publicized yet nonetheless staggering. These demands for
electric power in the United States double every ten years
thus increasing at a faster rate than the population, the Gross
National Product, the Research and Development Budget, the supply of scientific manpower, or almost any other measure of
growth in our affluent society. We can comprehend the immense
proportions of the power production task by considering that,
in the next decade, "we must add as much new generating capacity
as has been constructed since the invention of the light bulb.
If the increase continues at the present rate, the same amount
of capacity--as much as has been contructed through 1969--would
then have to be added in the following five years."[12] This
translates into an estimate of future loads that will require
over one billion kilowatts (electric) of installed generating
capacity by the year 1990.

Needless to say, the fulfillment of these requirements
will place sizable burdens upon the resource base of our economy
and, indeed, that of the world. First, if present trends continue for the next fifteen years, we will need approximately
67 per cent more oil, 33 per cent more coal, and 100 per cent
more natural gas than we have consumed to date,[13] while continuing to deplete our stockpiles of fissionable materials. Second,
the construction of large power plants is an extremely expensive affair ($200 million for a new nuclear plant) requiring
large capital expenditures. In 1960 utility companies accounted
for 20 per cent of all new U.S. corporate bond financing, and
each year 80 per cent of the industry's new money needs comes
from the bond market.[14] The high money rates of recent times,
taken together with rising tax burdens for private utilities,
have increased the carrying charges on power plant investments,
placing a strain on future earnings and causing some companies
to seek price increases from their regulatory commissions.
Third, the site requirements for large generating stations entail the purchase of several hundred acres of land, frequently
near heavily-populated metropolitan areas where land is at a
premium. Estimates of future demands indicate that over 250 new

plants will be required by the year 1990.[15] Fourth, power production places severe demands upon our environmental resources
of air and water. It is estimated[16] that power stations burning
fossil fuels (coal, oil, natural gas) are responsible for one
half of the sulfur dioxide and one quarter of the nitrogen
oxides that contaminate our nation's air. The air-pollution
problem is compounded by the fact that it is most economical,
in the conventional sense, to locate generating plants as
close as possible to the load; yet it is here, in the heavily-
populated, industrialized metropolitan areas that the air pollution is most severe. Another matter of great concern is
the effect of discharges of waste heat from nuclear power facilities to local cooling water supplies. Such plants operate
at thermal efficiencies much lower than those of fossil plants,
thereby producing more serious temperature increases in cooling
waters. It is estimated[17] that, by 1980, the electric power
industry will require the equivalent of about one sixth of the
total available fresh-water runoff in the entire nation for
cooling purposes. While cooling towers and cooling ponds are
technically feasible, they can involve cost increases of up to
20 per cent of the capital cost of an installed generating
plant, an extremely undesirable (albeit necessary in some cases)
additional economic burden. Apprehension over the impact of
the resulting temperature increases imposed upon bodies of
water whose life-sustaining capacity is more often than not
already badly weakened by other pollutants (sewage, industrial
wastes) has led to both national and local restrictive legislation.

     To meet the projected demands of the future, the techno-
logy of nuclear generating plants has forged ahead rapidly
since their introduction to the power market in 1964. Currently
a number of 1,000 megawatt (electric) plants are under construction, while it is the general consensus that the 2500 megawatt
(electric) liquid-metal-cooled fast-breeder reactor will replace present light-water units by the 1990's. Clearly, the ad-
vent of nuclear technology has been a major contributing factor

in the emergence of our nation into what is commonly known as
the "space age." Yet, for all our technological capabilities,
there is doubt for the first time that the electric industry's
capacity will be able to keep up with the rapidly escalating
per capita demands of our population, which is growing both in
numbers and in wealth. Each summer evidence accumulates indicating that the power companies are hard pressed to keep up
with these demands, particularly during peak hours. Many large
electric companies in cities of the Northeast, notably New York
City, have experienced "brownouts" while their appeals to customers
to reduce consumption during peak hours are becoming commonplace.
The occurrence of "brownouts" and "blackouts" poses a serious
threat to the health, safety, and well-being of the entire nation.
Glenn T. Seaborg, chairman of the United States Atomic Energy
Commission, has speculated on the possible outcries of angry
citizens "who find that power failures due to lack of sufficient
generating capacity to meet peak loads have plunged them into
prolonged blackouts--not mere minutes of inconvenience, but
hours, perhaps days, when their health and well-being, and that
of their families, may be seriously endangered. The environment
of a city whose life's energy has been cut--whose transportation
and communications are dead, in which medical and police help
cannot be had, and where food spoils and people stifle or shiver
while imprisoned in stalled subways or darkened skyscrapers--
all this also represents a dangerous environment that we must
anticipate and work to avoid."[18]

One might ask how the present situation has come about.
The answer is that a combination of unanticipated circumstances
has "handcuffed" the electric power industry to the point where
it is difficult for them to take the necessary steps to alleviate the pressures placed on them by increasing demands. The
two major stumbling blocks encountered by the utility industry
have been (1) long delays in construction scheduling, and (2)
difficulties in securing approval of site selections for new
generating facilities.

Delays in Construction Scheduling

A major problem facing the power industry is the continual slippage in construction schedules and escalation of site labor costs. This is usually associated with the specialized nature of the work involved in the traditional practice of constructing power stations as one-of-a-kind entities. For each new facility, a different set of laborers must be recruited, trained, and organized to carry out the specialized construction peculiar to that particular plant and geographic site. This problem is particularly acute in the construction of nuclear power stations which must be designed to meet stringent radiation containment standards, even in the event of severe seismic disturbances. Although factory-type construction techniques have been utilized in other types of large-scale construction, power stations are still built using traditional methods. As a result, we are witnessing inordinate delays in construction schedules: nuclear plants now take from five to seven years for completion.[19] These setbacks in construction scheduling are extremely costly and could wipe out any cost advantage that one particular type of plant might have over another. Also, power companies may be forced to anticipate delays by increasing construction lead-times, running the risk of premature retirement of existing facilities if the construction schedules are met. Also, if nuclear power is to be increasingly relied upon in the future, the installation of nuclear stations must proceed even faster than net power demand. The doubling time for nuclear plant capacity would then be on the order of seven years, since new demands must be satisfied while old fossil plants are phased out. Hence, the combination of short doubling time and long construction time could be a major obstacle in the way of increased reliance on nuclear power.

Difficulties in Securing Approval of Site Selections

The most severe problems facing the utilities of late have been associated with the selection of sites for new generating facilities. The situation has been accurately described as follows:[20]

> Everyone agrees that electric power supply is vital to
> the Nation and that we must find sites for the power
> plants needed to meet the Nation's rapidly expanding
> use of electricity. Nevertheless, "Don't Put It Here"
> is increasingly becoming the public's reaction to
> particular sites selected by the utilities. Further-
> more, the electric utilities are facing increasing com-
> petition for sites because our land resources are limited
> and the ingredients of a prime site for electric gene-
> ration also make it attractive to many other expanding
> industries.

This statement points out the two major difficulties related to site selection and approval--*competition* from a wide
range of prospective users, and the multiple pressures of *public
opinion*.

Competition for prime sites is not restricted to industrial development. The site that is ideal for electric power
generation is often very well suited for various forms of residential development, the location of transportation corridors,
commercial development, or recreation. This competition becomes
especially intense as the utility companies move to acquire
coastal locations to assure adequate supplies of cooling water.
Yet, as Senator Henry Jackson (D-Washington) pointed out in
his introduction in the Congress of the National Land Use Policy
Act,[21] many of these areas, "with the benefit of planning and
foresight, should have been reserved for other uses" such as
recreation, parks, or wildlife preservation. Strong arguments
of this kind have been made (see Chapter 3) as to the need for
preserving coastal resources in recognition of their extremely
high intrinsic value for recreation, conservation, and wildlife
preservation. The present trends toward locating power plants
at coastal and estuarine sites is in direct and irreversible
conflict with considerations of this sort. The State of California
has already located 85 per cent of its power stations on tidal
waters. Of large nuclear units now planned, built, or operated
in the United States, 18 per cent use ocean or bay water as
condenser cooling water and another 12 per cent are sited on

estuaries.[22] If this trend is allowed to continue for the next
20-30 years, 80 per cent of the cooling water for states bordering
on the Pacific Coast and 50 per cent on the Atlantic Coast will
be saline. Even if the ecological and esthetic effects of these
plants on the fragile marine environment can be demonstrated
as negligible, the use of large blocks of coastal acreage for
power plant siting constitutes a loss which cannot be regained
for use by future generations. Careful consideration of this
issue is of crucial importance in the formulation of long-range
planning for land-use management. In addition to market factors,
the utilities are likely to encounter increasingly stringent
constraints on site selection imposed by public agencies, such
as the conservative site standards set by the Atomic Energy
Commission with regard to areas of potential earthquake hazards.
All these factors are further compounded by the fact that the
greatest percentage of future sites is likely to be required
in the regions of heaviest concentration of population and existing
plant sites, especially in the Northeast Corridor. It is here
that land is the scarcest, especially at the seacoast.

The problem of public acceptance is primarily one of an
overriding concern for the quality of environment. The areas
of most concern are: (1) the air pollution caused by fossil-
fired plants; (2) the added thermal pollution caused by present-
day nuclear plants; (3) potential radiation hazards related
to nuclear plant operations; and (4) the visual intrusion of
generating facilities on the beauty of the natural landscape,
and other esthetic considerations. Presently, political action
has led to the situation whereby 20 per cent of new plants are
delayed by actual litigation, while 40 per cent are delayed
by general conservation and environmental considerations.[23]
It is reasonable to expect that problems of this general nature
will occur more and more frequently, causing delays of increasing
consequence when considered together with the delays in construction
scheduling.

A most recent manifestation of these multiple problems

associated with power plant siting can be found in a report
entitled "The Turkey Point Case, Power Development in South
Florida - A Study in Frustration."[24]   In this article, Harris B.
Stewart, Jr., Director of the National Oceanic and Atmospheric
Administration's Atlantic Oceanographic and Meteorological
Laboratory in Miami, documents an extraordinary chronology of
events (covering a *seven-year* time span) concerning the location
of a new nuclear power plant to satisfy the increasing demands
for power of the residents of Dade County, Florida. Turkey
Point was about the last remaining section of waterfront in
Dade County available for the needed expansion, a site which was
relatively remote from population (25 miles south of Miami and
5 miles from the nearest dwelling), was accessible to cooling
water (Card Sound to Biscayne Bay) and the transportation necessary to supply fuel for the units. Yet for the last seven years
the Florida Power and Light Company, despite evidence of good
faith on environmental issues, has been frustrated at every
turn in its attempts to secure approval of its expansionary
plans, with the most intense pressures coming from conservationists
who feared that the thermal effects of discharged cooling water
might be detrimental to the ecological systems of the area.
The issue has risen to national prominence, while at present
construction is at a standstill as the fight goes on in federal
courts. In concluding his examination of the conflict, Stewart
reflected on the dilemma:

> My personal feeling...is that the real endangered
> species in the overall ecosystem is man himself. Those
> who now scream that Biscayne Bay is being ruined by the
> warm water will be the first to rail against the power
> company when a power brownout or blackout occurs. The
> problem then is one of the conflicting uses of a re-
> source held in common--in this case the estuaries of
> our coastal zone. At the very heart of the problem of
> coastal zone management lies conflict between those who
> would use the waters for the cooling of electrical power
> generating plants and those who would keep our estuaries
> in their pristine, pre-man, condition. Some mutually
> agreeable meeting ground must be reached. It must not be
> considered as a case of power *or* estuaries, but rather a
> case of how to develop the power we require and still
> have estuaries that are needed for the development of

fish and for the many uses to which our growing coastal
population wishes to put them.

Certainly this constitutes an accurate expression of the central issue at stake; yet more and more we are finding, to our
dismay, that the multiple uses which we would like to see supported by our natural environmental systems are so incompat-
ible that use for one purpose often must necessarily preclude
use for many others. When we are confronted with basic dilem-
mas of this sort, we can only make decisions based on the relative weights of perceived value judgments of society as a whole.

We can now summarize the primary land-use issue regarding
the siting of electric power plants: we are running out of
usable inland fresh-water cooling capacity; we are running out
of coastal and estuarine land resources for recreation and conservation (let alone for power-plant siting); and we are running
out of patience with regard to the harmful side effects that
power generation imposes upon our environment. Perhaps it is
with the words "we are running out" that we can begin every
discussion of the allocation of our precious environmental resources.

So, having examined these difficulties facing the power industry, it is small wonder that we are now facing the prospect
of serious power shortages. The most immediate issue we face
today is not one of reducing high power costs or of choosing
between particular methods of generation (although this is certainly of great importance from a resource-consumption standpoint and other long-term considerations of national concern),
"but the vital one of persuading the American people that a
crisis exists right now"[25] in satisfying the *present* needs of
our highly power-dependent society. Sometime in the near future
drastic new approaches must be taken to alleviate the "saturation" problem (of which power production is an integral part)
before we exhaust our technological capabilities to hold back
disaster, before we exceed the ability of our environmental resources to disperse waste, and before we run out of usable land
for recreation or power-plant siting. It is inevitable that

we face in the long-run some serious tradeoffs (such as that
between undisturbed estuaries and power-generating plants, as
described by Mr. Stewart) given that we continue with our
present patterns of exponential growth in so many areas. Perhaps we must eventually try to cut back on consumption by
buying fewer air conditioners, television sets, and cars, although this would seem socially unacceptable in our present
democracy. The only *really* effective long-run solution to all
congestion-related problems is to attack the source of conges-
tion--continued population growth. Yet solutions are *also* needed
in the short-run; the vanguard of crisis is here and now. We
must find measures to avert dangerous power shortages with an
eye to the future consequences of our actions on man's total
environment. Perhaps we are not adequately equipped at present
with the institutional mechanisms (social, economic, political)
to completely resolve issues of long-term significance brought
on by the preponderance of man's presence on this earth. Until
we become so equipped, we can turn to technology to provide the
short-run solutions to pressing problems such as the one here
described--realizing at all times that we are just buying time
and that the consequences of failure at some later time may be
all the more severe.

2.   The New England Situation

In New England, the reason for special concern is clear.
It is projected that over the next twenty years or so the
power system of the Northeast Corridor will be approximately
3.5 times its present size.[26] The problems of air pollution in
most large cities of this region are partially attributable
to the relatively exclusive use of fossil fuels for power generation. In 1969, 82 per cent of all the power generated in New
England was produced by fossil-fueled plants.[27] Increased reliance on nuclear generation has run into the situation where
usable inland fresh-water cooling capacity is already all but
exhausted in terms of allowable ecological margins. While cooling towers and ponds are technically feasible, they involve cost

increases of up to 20 per cent of the capital cost of an installed generating plant, an extremely undesirable additional
burden in view of the chronic high power costs in the region.
For nuclear power, the possible cost reduction advantages of
cheaper fuel would be negated by pollution control expenditures
of this sort. Table 2.1 shows a comparison between New England
costs and the composite average of all regional costs in the
United States. Note that, while the additional cost of cooling

Av. Price (Revenue/Sales)	U.S. Composite (mills/Kwhr)	New England (mills/Kwhr
Residential	20.9	26.4
Commercial (large light and power)	9.1	14.1
Estimated Busbar Price (new fossil-fired plants)	7.16*	7.99*
Estimated Busbar Price (new nuclear plants)	6.85*	6.88*

*These figures are estimates based on calculations described
in Appendix A.

Source: Edison Electric Institute, Statistical Yearbook of the
Electric Utilities Industry for 1969, New York (1970)

Table 2.1 New England Electricity Prices - 1969

apparatus does not substantially affect residential rates** (on
a percentage basis), it can have a substantial effect on large
commercial and industrial consumers, who account for about 35

**Most of the cost of electricity to New England residential con-
sumers (26 mills/Kwhr) consists of transmission and distribution costs. Large industries can substantially reduce these
costs by locating near the generator (busbar) where the average
cost is close to 7 mills/Kwhr (for new plants), and by using
power in large block amounts. (See Table 2.1.)

per cent of the total sales in New England each year, and who
are already paying abnormally high prices (due in part to the
high average cost of fossil fuel in this region). Here then
is the basic dilemma facing the New England region in the field
of power production: do we relieve the environmental stress
by increasing the price of electricity (at the expense of further debilitating effects on the economic posture of regional
industries) or by forcing the power industry to absorb the
costs of additional measures for environmental protection (even
though they are currently facing crises in capital financing,
construction scheduling, and site approval)?

3.   The Proposed Solution

A recently-evolving trend places increased emphasis on the
use of ocean water for cooling purposes. There is a certain
amount of conventional wisdom and precedent in this solution.
The State of California, which shares at least superficially
in some of the same power-production problems, has already located 85 per cent of its power stations on tidal waters.[28]
The present study has led us to the conclusion that the use of
ocean water for cooling is a basically sound concept, but sadly
deficient in the way in which it is being implemented; namely,
through construction of seaside power plants. While adequate
cooling water supplies would be available, all of the major
problems (as discussed above) would remain, with pollution and
land use presenting the most serious difficulties. The land-
sea interface is a particularly vulnerable ecozone whose shallow
waters might still be susceptible to the harmful effects of
heated effluent from a nuclear power plant. Furthermore, as
the study reported in Chapter 3 testifies, the shoreline has an
extremely high intrinsic value for other competing uses such
as recreation and esthetic and ecological preservation.

Thus the heart of the problem, as finally formulated,
appears to lie in how best to implement the use of the ocean as
a heat sink for the effluent discharges associated with electric
power production *without* encountering the economic, social, and

political stumbling blocks in proposals involving land-based
plants. It is in this context that we present a concept that
strikes at the heart of the power-production issue, providing
relief for pressing problems and holding great promise for the
future. This concept has the relatively unique feature that
it has the potential for simultaneous solution of the two major
difficulties faced by the power industry--construction delays
and site selection and approval--without placing additional
stress on the financing aspects of the overall situation, while
removing a serious area of contention and conflict from the
already overburdened shoulders of land-use planners. The concept is set forth and examined in the remaining sections of
this chapter.

III. THE OFFSHORE CONCEPT

In recent years the concept of locating large electric
generating stations at offshore sites has gained increasing
attention. There are a number of extremely attractive aspects
to this concept. One is the potential elimination of many of
the difficulties associated with the selection and approval of
sites. Questions of land cost and availability are no longer
relevant; competition with industrial and other development interests would be nonexistent; use of the ocean's capacity for
cooling seems to be the only answer to the environmental problems of thermal pollution; and siting of plants offshore allows
new flexibility in locating close to the load, especially as
our population concentration shifts to the coastal perimeters
of the nation where land is already at a premium. Hence,
nearly all the problems of land-use management associated with
the siting of power-generating facilities can be effectively
obviated.

The second major advantage of great importance is the
amenability of many offshore designs to shipyard construction.
While U.S. shipyards are presently operating with a backlog of
orders, their utilization is subject to large variations depen-

ding on the construction plans of the Navy and other military
customers. This is because the U.S. yards are generally not
competitive on the world market; the costs of a given shipyard
product are approximately 20 per cent less in Europe and 35 per
cent less in Japan where shipyards are larger. Diversion of American shipyards to the mass construction of power plants might constitute a more efficient use of this well-developed resource
while having beneficial side effects on regional economies.
Shipyards are geared to hold tight construction schedules. For
example, it is estimated[29] that large (43,000-ton displacement)
nuclear-powered containerships could be produced in as short a
time as 18 months at a shipyard price on the order of 40 million
dollars. Compared to these ships, a power plant is a very
high-value product: a 1,000 megawatt electric plant now
costs a utility over 200 million dollars. Furthermore, like
ships, power plants could become an important regional export
industry. The market for the Eastern and Gulf Coasts of the
United States alone is estimated to average 10 plants per
year,[30] worth a total of more than two billion dollars for the
next several decades.

Another benefit, perhaps the most important, is that the
construction of a power plant at a shipyard has the potential
for significant savings in construction time. A shipyard maintains the permanent base of shops, equipment, and skilled labor
that is lacking in the traditional methods of power-plant construction. This provides for (1) increased stability of the
skilled labor force, (2) increases in the efficiency of the
skilled labor force by allowing a *learning curve* to develop as
additional stations are built, and (3) elimination of duplica-
tive areas of management, management support, engineering, construction support, and quality control presently necessitated
by separate construction locations. Shipyards are also frequently hubs of transportation networks that can use the most
efficient combinations of land, sea, and air facilities to reduce transportation costs over on-site construction of land-
based plants. The possibility of shortened construction times

to help the power industry stay ahead of rapidly-increasing consumer demands, together with potentially large savings in
capital outlays by utility companies for new plants, are most
encouraging prospects in these times when substantial delays in
construction are both costly and commonplace. While shipyards
in other sections of the country could tool up to compete with
those in New England, it is clear that the combined assets of
this region give it a competitive edge due to a concentration
of established firms engaged in large-scale operations in shipyard construction and in power-plant design, construction, and
operation.

Several designs for offshore power stations have been proposed in the literature. These designs usually fall into two
broad categories: the basic choice is between *floating* plat-
forms or enclosures (indirectly coupled to the ocean floor by
a mooring system), and *fixed* structures that are solidly attached to the bottom.

Fixed Structures

Fixed structures can be of several forms:

    (1)   Man-made islands

    (2)   Fixed-pile platforms

    (3)   Jack-up platforms

    (4)   Grounded barge

(1)   Man-Made Islands

The technology of this scheme does not differ in any
appreciable way from that of land-based plants, except in the
added complexities (and cost) of site preparation and transportation of men and equipment across a water gap. Studies investigating the feasibility of this concept have been undertaken
on both the East[31] and West[32] Coasts.

(2)   Fixed-Pile Structures

In this scheme, piles are floated or barged to a given
location and erected permanently. A platform is then built

on these piles, similar to a Texas-tower configuration. Japanese engineers[33] have proposed construction of an offshore power station based on this design. Like the man-made island concept, this design entails conventional land-based construction techniques, again with the added complexities of offshore, on-site construction.

(3) Jack-Up Platform

The power station is constructed in a shipyard atop a platform equipped with extendable legs, floated to the chosen site, then jacked up on the legs (grounded to the seabed) out of the water and into a position similar to that of the fixed-pile structure. The jack-up system is generally thought to be useful in depths up to 250 feet, while 300 feet is accepted as the practical limit.

(4) Grounded Barge

This scheme involves the permanent grounding of a floating barge-like platform to the ocean floor (in shallow water) or on a prepared site. The barge and power station are again constructed in a shipyard, towed into position and then ballasted until the grounding is complete.

Floating Structures

A number of designs have also been suggested for floating structures. These include:

(1) Submersible stations
(2) Ship hulls
(3) Barges

(1) Submersible Stations

This concept is presently under investigation by the Electric Boat Division of General Dynamics under the auspices of the Department of the Interior.[34] As described in a proposal by R. W. Marble,[35] this scheme consists of a grouping of cylindrical containers enclosing the reactor, steam and electrical systems, positioned at sea and tethered to the bottom by a multi-

point mooring. The position of the station with respect to the
ocean surface would be controlled by ballasting in much the same
way as on a present-day submarine. The best position for the
station with respect to the surface must be ascertained--at
least 100 feet over the main hull section probably would be desirable. In shallow water this might demand bottom siting.
A personnel transfer system would be provided by a long access
trunk and elevator to a heliport structure above sea level.

   (2) Ship Hulls

   The U.S. Army presently operates the STURGIS,[36]
a floating nuclear power station utilizing a small pressurized-
water reactor (PWR) in a conventional ship hull. For the much
larger plant sizes necessary for commercial power generation,
this shape would not be suitable because of excessive stability
problems and the difficulties in designing a suitable mooring,
although the good towing characteristics would provide excellent
mobility for smaller-size plants. Experience with such reactors
aboard a number of naval vessels, including the N.S. Savannah[37]
and the Otto Hahn,[38] have indicated that "there are no inherent
reasons why reactors should not be installed on floating plat-
forms--at least not for pressurized-water reactors. Some concern has been expressed in the past about the performance of
boiling water reactors on ships...under conditions of roll,
pitch, and heave. Several studies, plus the performance of the
Otto Hahn, suggest that this is not a serious problem."[39]

   (3) Barge-Mounted Power Stations (BMPS)

   This concept involves the location of a power plant
on a floating barge, built in shipyards and towed to and moored
at appropriate offshore locations. One possible configuration
that has been suggested in a recent study[40] consists of a rec-
tangular barge supporting the reactor and its containment dome
in the center, surrounded by the Personnel and Generator
housings. Access for operating and maintenance personnel could
be provided by a causeway, while docking facilities could be
constructed if boat transportation proved more economically or

technically attractive.

It should be noted at this point that determination of the
technological feasibility of the offshore concept should not be
restricted to consideration of nuclear plants alone. In many
respects, the issues are independent of the type of power plant
that is located offshore. It may be that locating fossil-fired
plants offshore might be desirable if the atmospheric dispersion
conditions are favorable enough to avoid air pollution in nearby
cities. This report has based the analysis on nuclear technology since the literature related to the offshore concept has
been developed primarily in this area and because of the projected
dominant role of nuclear power in satisfying future energy needs.

The literature describing variations of the offshore concept has focused primarily on three designs: man-made islands,
submersible stations, and floating barge-mounted facilities.
The general consensus is that all of these concepts are technologically feasible,[41] the ultimate determinant being that
of cost. Of the three, the man-made island concept appears to
be the least desirable. While this alternative avoids many of
the difficulties associated with site selection and approval,
the economics of construction may well be prohibitive, as was
the case with the previously-referred-to BOLSA island project
in California. We have seen in the foregoing analysis that we
must be particularly sensitive in formulating our solution to
the question of construction scheduling. Considerations of this
sort lead to the rejection of the artificial island concept as
it entails not only the added time necessary to construct the
island itself, but also the additional time and costs required
to transport all labor, equipment, and material across a water
gap--a formidable task in itself. Similar arguments can be made
against fixed-pile structures. On the other hand, the floating
and submersible designs each have singularly attractive features. For example, both designs appear to provide near-
absolute protection from seismic disturbances.[42,43] In addition, it is thought[44,45] that underwater containment of nuclear

reactors would provide much better post-accident fission-product retention than is now possible in land-based plants. Also, floating stations offer a relatively limitless choice of locations since they would be unaffected by ocean depth or bottom contour. The possibility of moving stations at some future time in response to changes in population or consumption patterns might be a factor in favor of floating stations. Having carefully weighed these and other factors, we conclude that the Barge-Mounted Power Station (BMPS) appears to have the most potentially attractive aspects of all the alternative offshore designs. Hence, the BMPS has been selected as the reference design for the study reported in this chapter.

It is clear at this point that the offshore concept contains a number of extremely attractive features that go a long way toward eliminating the twofold problems faced by the power industry--costly construction schedules and difficulties with site selection and approval. If the technology and the economics of this proposal do not provide new obstacles of complexity comparable to today's problems, the realization of the offshore concept could be one of the most significant advances in the history of electric power production. Thus, the remaining sections of this chapter are devoted to a more detailed analysis of the technical feasibility and economic viability of the BMPS concept.

IV. TECHNOLOGICAL CONSIDERATIONS

While questions of technological feasibility and cost are closely linked, it is useful for present purposes to separate them. The economic analysis, discussed in Section V, is based mainly on a plant size of 1,000 megawatts electric (MWE), since this is the size of present new plants and because reliable cost information is therefore available. On the other hand, the technological feasibility is assessed on the basis of a 2,500 MWE nuclear plant, the unit size projected for the 1990's. We adopt this difference in viewpoint because it is quite germane

to the question of technological feasibility whether future
designs can be accommodated by the BMPS concept. The 2,500 MWE
plant is a factor of approximately 25 larger in output than the
propulsion plant for the proposed nuclear-powered containership
cited previously, and 250 times as large as the only existing
nuclear BMPS, the U.S. Army's STURGIS.[46] Furthermore, we have
investigated the technical feasibility, assuming that the reactor type would be a liquid-metal-cooled fast-breeder reactor
(LMFBR), since it is the general consensus[47] that this type
will supersede present light-water types in the 1990's.

1. General Features of the BMPS

In beginning this study of offshore power-plant siting, we
were fortunate to have access to information developed by
R. W. Marble of the Electric Boat Division of General Dynamics
concerning a submerged offshore power station concept.[48] The
main advantage that this submersible plant would have over a
surface plant would be greater insensitivity to storm, wind
and wave action and a lower collision probability with ambient
shipping. Consultation with members of the M.I.T. Department of
Naval Architecture and Marine Engineering indicated,[49] however,
that it was a quite reasonable expectation that a suitable
mooring scheme could be developed for a barge. The barge scheme
was then adopted as a reference design because of its advantages
in terms of less complex and less costly design, and easier
accessibility.

The general features of a BMPS are sketched in Figure 2.1.
An enclosed barge houses the entire plant, the enclosure serving
as a secondary containment vessel for the nuclear station. On
station the barge is negatively ballasted and partially awash in
order to maximize stability. Several mooring schemes are possible; that shown in Figure 2.1 consists of extendable legs
with concrete feet. Under tow the legs are raised; on site the
legs are lowered until the feet rest on the (prepared) bottom;
then the barge is ballasted to partially sink down onto the
legs. Although the entire question of mooring requires further

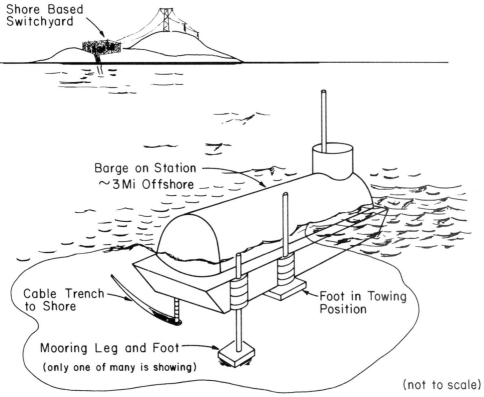

Figure 2.1   Artist's Conception of a BMPS Moored on Station

investigation, and ultimately model testing, there is sufficient experience[50] over the past several decades with semisubmer-
sible oil rigs to give assurance that the mooring problem is
soluble.

A second important question concerns the size of the barge
required. A preliminary plant layout and weight study indicated
that a barge 950 feet long, 120 feet wide, with a towed draft of
30 feet and having a towed displacement of 110,000 tons would
be required. While larger than barges now under construction
(the largest described in the literature were a 532 ft x 87 ft
petroleum barge,[51] and the 400 ft x 100 ft AGATTU deck cargo
barge[52]), it is not particularly large in comparison with the
hulls of large containerships and giant supertankers, as shown
in Table 2.2. Consultation with Mr. Les Stypinski of General
Dynamics, Quincy Electric Boat Division, has indicated that barge
hull construction within this size range would be feasible.

	BMPS	LARGE CONTAINERSHIP*	SUPERTANKER**
Length (ft)	950	900	1,100
Width (ft)	120	120	180
Draft (ft)	30	30	60
Displacement (tons)	110,000	55,000	3326,000
Approx. Cost (million $)	200	35	30

*Sources: World Bulk Carriers, Fearnley & Egers Chartering
           Co., LTD. January 1970.

           The Bulk Carrier Register, H. Clarkson & Co.,
           LTD. (1970).

**Sources: Ocean Industry, Vol. 5, No. 1, p. 35, January 1970
            and Vol. 5, No. 12, p. 18, December 1970.

Table 2.2. Comparison of BMPS with Large Commercial Vessels

2. Design Characteristics

Figure 2.2 shows the schematic cross section of a barge-mounted power station. The barge is divided into three regions: the so-called Personnel, Reactor, and Generator Spaces. The last of these contains the turbine, generator, condenser and condensate system, and transformers for high-voltage transmission. The switchgear and other auxiliary equipment are located on shore. The generator part of the plant does not differ in any substantial way, other than in terms of a somewhat more compact arrangement, from modern steam plants now used in conjunction with fossil-fueled stations. While it is likely that

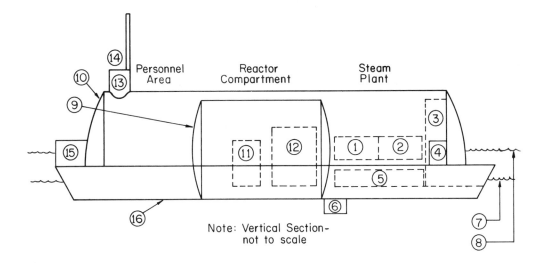

Figure 2.2   Schematic Layout of Barge-Mounted Power Station

single turbo-generator sets having a 2,500 MWE capacity will be available in the 1990's, the arrangement shown in Figure 2.2 uses two 1,250 MWE units in parallel. This decision was based on the consideration that reliable size and weight data are now available on 1,250 MWE units, and it may well prove preferable to have greater reliability through redundancy in these large plants.

The Personnel Area contains the reactor and steam plant
control stations, administrative offices, and auxiliary equipment compartments. It will differ from a land-based station
primarily because of the need for providing more complete facilities for comfort of personnel in situations where they cannot
reach shore for extended periods due to severe weather conditions. If the plant is to be located close enough to shore so
that a causeway can be built, such provisions would be unneces-
sary.

The reactor compartment is located amidships because of
stability considerations related to its greater density as compared to the rest of the plant, and because this location has
maximum protection in case of a collision. The reactor is
housed inside a primary containment vessel, which in turn is
located inside the enclosed barge that acts as a secondary
containment vessel. In this respect the BMPS has a more conservative containment arrangement than present land-based stations.
The secondary containment vessel is maintained at a slight vacuum relative to the external atmosphere, and the primary containment atmosphere is at an even lower pressure. Thus, during normal operation, all leakage is inward, and during a severe accident the primary vessel can relieve to the secondary containment
space if overpressurized. This concept is very similar to the
separate vacuum building method developed by the Canadians. By
designing the barge in three distinct sections one also enhances
the capability for simultaneous piecewise construction, a common
shipyard practice.

The reactor type chosen for this study is the liquid-metal-
cooled fast-breeder reactor having parameters scaled up from
recent AEC-sponsored 1,000 MWE design studies.[53] It is important to note, however, that the BMPS concept is equally applicable to present-day pressurized or boiling water reactors, to
fossil-fueled stations, or even to advanced concepts such as
magnetohydrodynamic (MHD) generators. There are some inter-

esting aspects pertinent to the use of the BMPS concept in conjunction with fossil-fueled stations which deserve mention.
The first important point is that significant cost reductions
might be possible. By locating the power station closer to deep
water it will be possible to use giant supertankers for fuel
delivery, thereby substantially reducing fuel costs. Storage
tank costs can also be reduced by adopting the large submerged
tank concepts now being planned for offshore tanker on-loading
facilities at the oil production fields in the Mideast. Ecological advantages also enter in. By keeping the tankers out of
congested harbors and shallow water, the probability of severe
accidents (as have occurred notably in San Francisco Harbor)
leading to extensive oil spills is reduced. Offshore siting of
the plants could also take advantage of prevailing westerly
winds to keep air pollutants away from land. This latter advantage can be traded off for further cost reductions by reducing
or eliminating the need for anti-pollution measures and devices
such as the use of low-sulfur fuel or installation of stack-gas
cleanup systems.

Since the detailed features of the generator plant, either
fossil- or nuclear-fueled, are not particularly germane to this
analysis, they will not be discussed here further.

3. Engineering and Operational Issues

There are a number of issues that must be resolved as a
prerequisite to successful realization of the BMPS concept.
These issues are outlined and discussed only briefly here; for
a more detailed analysis of the major technological, economic,
political, and legal aspects of the offshore concept, the reader
is referred to a follow-up study[54] conducted at the Massachusetts
Institute of Technology during Fall 1970.

First of all, the successful dispersion (in an ecologically-
safe manner) of heated water discharged as condenser effluent
must be demonstrated. This includes a consideration of the effects on the local ocean environment (bottom flora, fauna, etc.)

and on the ecosystem of the shoreline. One possible dispersion
scheme is to use stratification analogous to a normal summer
thermocline. The intake for condenser cooling water would be
located at about mid-ocean depth directly under the structure,
and the discharge of warm effluent would be at the surface.
The density difference will cause the effluent to remain floating
on the surface, thereby preventing recirculation. This scheme
seems ecologically safe because it simulates normal summer solar
heating in keeping warm water away from the sensitive flora and
fauna on the bottom.

It is a complex problem to analyze the effects on the shoreline of the discharge plume from an offshore plant, complicated
primarily by the combined effects of the tides and ocean currents at each particular location under study. The combination
of ocean currents that usually parallel the shore and tidal
motions would eliminate the possibility of stagnant effects as
well as prevent the discharge from moving directly toward the
shore. While the ultimate ecological determinants of how close
to shore the plant could get are based on the particular oceanographic characteristics and legally allowable temperature increases of the area, preliminary figures[55] seem to indicate that
a site that is at least one mile from shore in at least 25 feet
of water should have acceptable ecological effects. This of
course is subject to legal constraints imposed by local political bodies with jurisdiction over whatever shoreline areas are
involved.

Any offshore power plant must be protected against collision damage. The use of buoys, radar, or lighthouse beacons can
help warn shipping to avoid the site. Submarine nets or floating raft structures can also provide protection. In addition,
a collision barrier consisting of layers of edge-on plates can
be constructed between the outer hull (secondary containment
vessel) and the inner primary reactor containment. If break-
waters are used for wave dissipation at shallower sites, these
could also provide a good measure of collision protection.

Even if a ship were to penetrate all collision protection devices, a barge with the reactor in a central location could suffer the loss of several outside compartments without impairing
safe reactor shutdown. The largest difficulty in this area
deals with the possibility of a loss of mooring. Steps must
be taken to provide backup mooring systems for the floating installation in the event of a disruption of the primary system.
Also, some careful consideration of the probability of airborne
collision should be included in the detailed analysis, since the
reactor dome may be as much as 150-175 feet above sea level.

To protect the plant against extreme environmental effects
such as 30-foot storm waves, 7-foot tidal waves, and 200-mile-
per-hour winds, a rock breakwater can be economically constructed
in depths less than 50 feet. This breakwater can also serve as
the basis for the mooring system and as a collision shield for
the barge. There is evidence that excellent protection from
seismic disturbances would be provided by the fact that the barge
is floating, while "air springs" might be designed which would
help shield the barge from vertical shocks. All these issues are
more thoroughly discussed in a paper by Harold M. Busey of the
Donald W. Douglas Laboratories.[56]

Certainly one of the most important considerations related
to safety is that of radiation containment. In this area, two
sets of criteria are applicable. The first is containment following a reactor accident that is brought about by some malfunc-
tion within the plant itself. Standards for such occurrences
are well developed and would be incorporated in the basic design
of the plant with little or no modification necessary from the
design of a land-based installation. The second criterion is
that the containment system remains intact in the event of accidents due to external factors such as earthquakes, severe storms,
tidal waves, collisions, etc. A land-based station must be designed to withstand earth shaking and differential movement
under maximum credible seismic conditions without causing any
public hazard beyond that acceptable by current AEC standards.

The criterion that is often used is a maximum allowable displacement of six feet in any direction at an acceleration (load factor) of 0.7g.[57,58] This extreme condition is greater than that ever measured for an earthquake and can be used as the criterion for calculations of offshore seismic protection provided by the different alternatives. Since offshore stations can probably meet the above criteria, then presently available containment vessels would be adequate for direct application to the offshore concept.

One operational problem that arises is that sea-based plants will face greater corrosion problems. This requires use of more expensive condensers. However, this problem is already faced by all other plants using tidal or ocean water for cooling and is demonstrably soluble. The remainder of the plant is protected by the outer barge hull; thus the only other major corrosion problem is that of the hull itself. Periodic docking and overhaul, as with ships, is impractical. However, very effective protective coatings are available. They have not been very successful on ships because of erosion due to ship motion-- a problem not encountered in a moored barge. In addition, no special poison paints would be necessary to prevent the formation of barnacles (which have always presented a troublesome maintenance problem).

The power generated by an offshore station could be transmitted to shore either by submarine cable or by overhead transmission lines. The choice is a matter of reliability and cost, since the necessary technology exists. Submarine cables seem to have the edge in reliability. At least one commercial firm has previously bid on 345 kV submerged cables. With this system three cables would be necessary, each cable being a triple conductor. Two cables are capable of carrying the entire load and one is reserved as a spare. Total cost would be approximately 3.0 million dollars for the first mile and 2.5 million dollars for each additional mile. Overhead transmission lines have not been built with towers directly in the water, but the cost would probably be less. Built on land, some lines have

spanned 4,900 feet. The Central Electricity Generating Board in
England has used 400 kV lines with towers 4,500 feet apart.
Such an arrangement might be suitable for plants less than a
mile from shore, while at greater distances towers must be built
in the water. This would decrease reliability in severe weather
or in case of collision with the tower.

The transmission issue presents another major trade-off
variable, mostly in terms of cost, providing a constraint on the
feasible distance from shore for an offshore plant. One major
engineering issue that must be confronted in this regard is the
design of a reliable interface between a dynamic barge and here-
tofore-rigid transmission facilities.

Another important factor in the offshore concept is the
amenability of various structures to shipyard production. The
advantages of this scheme have been outlined in general in previous discussion. The idea behind building the offshore power
station in a shipyard is to:

1)   Take advantage of existing facilities and capabilities
     for construction of the plant in a centralized area.

2)   Take advantage of a stable, skilled work force which
     has experience in constructing steel structures.

If the concept is feasible, then presumably there would be a
learning curve associated with the construction of each new plant
with potential for even greater savings in time and labor as
time goes on. The BMPS concept would have to demonstrate potential as a continuing profit-making enterprise before facilities
or practices of existing yards would be diverted to construction
of floating power installations.

There are certainly many other problems which have to be
resolved in preparing a final detailed analysis. Transportation
of personnel, the effects of winter icing, replacement of spent
fuel, and heavy maintenance all have aspects which differ from
shore-based precedents. In this brief review we have limited
discussion to major problems suggested by knowledgeable experts

who have reviewed the BMPS concept and to those aspects found
to be most pertinent within the context of our analysis.

4. Summary and Conclusions

To summarize, our treatment has led to the identification
of a number of major trade-off issues that must be assessed in
the choice among alternative schemes and their effect on costs.
These include:

(1) The degree of seismic protection required.

(2) The suitability of various mooring schemes and their
ability to withstand the forces of maximum storm conditions for a large number of years.

(3) The minimization of ecological stress on local sea
life and on coastal ecosystems to within allowable
margins (a function of oceanographic conditions and
constraints imposed by the institutional environment).

(4) The protection against collision with air and sea
vehicles.

(5) The effect of tidal movement and tsunami (tidal wave
triggered by a remote earthquake).

(6) The establishment of a thermocline to prevent mixing
cold influent with hot effluent--constraint on allowable depth of water at the site.

(7) Basic design considerations, e.g., stability, towing
characteristics, the limitations on draft of floating
structures imposed by the depth of harbors providing
shipyard access, etc.

(8) The suitability for use with undersea or tower transmission lines (cost is the major variable).

(9) Ease of access for personnel and suitability for
heavy maintenance.

(10) Radiation containment and radionucleid waste disper-
sal during normal and accident conditions--must conform to AEC standards.

(11) The adaptability of various offshore structures to shipyard construction without incurring prohibitive costs.

(12) The susceptibility of the plant to sinking and loss
      of access, and how this might interact with design
      features such as use of soluble poison control.

(13) Effect of winter icing on plant operations.

(14) Coupling to onshore transmission facilities and suit-
      ability for clustering to reduce costs: i.e., how
      far apart should multiple plants be?

(15) Refueling and maintenance schemes adopted: need for
      boat or truck access, etc.

The single major conclusion to be drawn from the foregoing
analysis of various aspects of the offshore concept is that *it
appears technologically feasible without undue (and, in most
areas, no) extrapolation of already-existing technology*. This
conclusion is supported by the results of some other studies of
the offshore concept, including those by H. G. Arnold, et al.,[59]
Daniel, et al.,[60] and H. M. Busey.[61] We can therefore turn at
this point to the question of *economic* viability as the ultimate
determinate of the feasibility of the offshore concept!

V.   ECONOMIC ANALYSIS

The success or failure of the BMPS approach depends in
large measure on the economic incentives that can be demonstrated in its favor. This section will attack the question of
economic viability in a very general manner. The procedure followed will be to compare the projected cost of a 1,000 megawatt
electric BMPS with that of other stations of the same size.
The 1,000 MWE size was chosen because this is the average unit
size of power plants now under construction and we can therefore
speak with some confidence about actual, not hypothetical, cost
figures. For the same reason, the primary emphasis will be on
nuclear reactors of the light-water type (pressurized or boiling
water reactors), since these monopolize the current U.S. commercial reactor market.

A very general approach to the question of economic viability proved possible for several important reasons. First of

all, the preliminary differential cost analysis shown in Table 2.3 comparing the BMPS to a land-based station shows that capital costs should be the same within the accuracy of the rough-cost estimates performed: approximately 200 million dollars for a 1,000 MWE plant.[62] Secondly, the largest single difference in final overall plant cost proved to be due to the compression in construction schedule possible with the BMPS. Since this saving is merely a consequence of the cost of borrowing money, the most important question of cost differential can be settled without resort to a discussion of any technical factors.

Category	Land-Based Costs (x 1000 dollars)	Barge-Mounted Costs (x 1000 dollars)
Structures and Improvements	21,200	14,200 (saving is on cost of barge)
Reactor Plant Equipment	72,400	72,400
Turbine Plant Equipment	52,700	52,700
Accessory Electric Equipment	7,000	10,000 (additional cost is for underwater cable)
Misc. Power Plant Equipment	1,900	1,900
Total Direct Cost	155,200	151,200
Indirect Land Costs	45,800	45,800
Total Construction Cost	200,000	196,000

Table 2.3 Preliminary Capital Cost Estimates for 1000 MWE Nuclear Power Plants

1. Savings Due to Shortening of Construction Period

Construction of a power plant at a shipyard has the potential of significant savings in construction time. A shipyard can provide a permanent base of skilled personnel and a variety of shops and heavy equipment, as opposed to current on-site construction practices in the power industry where every job is, in

effect, a one-of-a-kind effort started from scratch. Thus it
was found to be quite important to assess the effect of construction time on the cost of a plant and, hence, on the cost of the
power that it would produce.

   Table 2.4 summarizes the economic ground rules which are
representative of current utility accounting practices in studies
of this sort. The determination of savings due to shortened
construction time consists of the calculation and comparison of
the net present worth (investment) of the plant (as of the date
of startup) for different construction times. All plants were
assumed to start delivering electricity on the same calendar
date. The net capital investments can then be translated into
a fixed charge component using the method described in Appendix A.

Plant Size:	1,000 megawatts electric (MWE)
Capital Cost:	$200 million, nuclear $170 million, fossil
Lifetime:	30 years
Capital Structure:	60% bonds at 8% annual interest rate, 40% stocks at 13%
Construction Schedule:	S-curve (hyperbolic tangent)
Reference Construction Period:	5 years
Load Factor:	0.8
Fixed Charge Rate:	17.1%/yr
Operation and Maintenance Costs:	0.30 mills/Kwhr
Cost of Nuclear Insurance:	0.10 mills/Kwhr
Cost of Nuclear Fuel:	1.50 mills/Kwhr
Average Cost of Fossil Fuel (in New England):	3.50 mills/Kwhr

Table 2.4   Ground Rules for Economic Comparison

   The studies were carried out for an S-curve construction
payment schedule which is typical of current site construction.
However, calculations were also made using a linear schedule

which is typical of shipyard construction, and the difference
in total cost was found to be negligible.

Figure 2.3 shows the results of these calculations. For
reference purposes note that the average production cost of

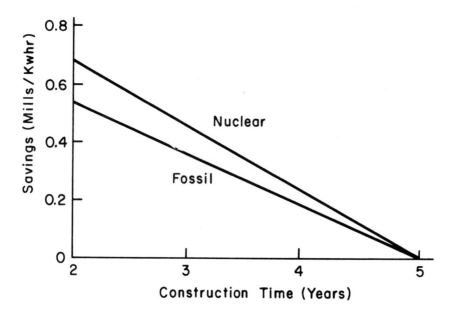

Figure 2.3   Savings in Cost of Electricity by Compression
Construction Schedules

electricity at the busbar (output of the switchyard) for a new
plant of this size would be around 7 mills/Kwhr (see Appendix A).
Thus, the savings shown in Figure 2.3 represent up to 10 per
cent of the total cost. Also note that 1 mill/Kwhr represents
around 6.8 million dollars per year for a 1,000 MWE plant operated at 80 per cent load factor.   This one feature alone--
the savings in the cost of borrowed money--represents a large
potential saving for the BMPS concept.

If it were not for other considerations, this item would
represent the major cost differential upon which a utility would
base its decision as to whether a BMPS should be constructed in

lieu of a land-based station. However, ecological costs (that
are relatively easy to estimate in this case since they include
capital costs for new equipment) should now be injected into the
picture; hence, a broader range of alternatives merits analysis.

2. Comparison of Alternatives

In order to take into account a sufficiently broad range
of alternatives, we have expanded the economic study to consider
a number of other concepts. First of all, we considered the
additional costs imposed due to pollution control. This represents the cost of cooling towers to alleviate thermal pollution
of water resources for both fossil-fueled and nuclear stations
constructed inland. Fossil-fueled plants are assigned an additional penalty to account for use of low-sulfur fuel and for
stack-gas cleanup to alleviate air pollution.

In addition, we have estimated the economic impact of future
developments. By the 1990's, plant sizes will have more than
doubled[63] due to economies of scale, and the fast-breeder reactor with its exceptionally low fuel cycle cost should be in commercial operation.

Tables 2.5 and 2.6 record the results of these cost estimates. All estimates show the price of electricity at the
busbar of the plant. As was previously pointed out, there is
little that can be done to reduce the costs to the residential
consumer since he pays mainly for transmission and distribution
and we are concerned with the cost of generation. However, there
can be a substantial cost reduction to the large industrial consumer who pays a price much closer to the busbar cost. Indirectly, these savings are also felt by the average consumer
when he purchases the products of New England industry.

The first three rows of Table 2.6 compare the costs for a
nuclear station constructed on land with BMPS construction
schedules of four and three years respectively. Since the BMPS
is practically unaffected by pollution control costs, it entails
a substantial savings of up to 1.45/Kwhr over the land-based

design.

| FACILITY | CONSTR. TIME(yrs) | NEW PLANT EXPENDITURES | | | | POLLUTION COSTS | | TOTAL |
		Capital Costs	Oper.& Maint.	Nuclear Ins.	Fuel	Air	Thermal	
Nuclear Land-Based	5	4.95	.30	.10	1.50	1.10		8.05
Nuclear BMPS	2	4.17	.60	.10	1.50	–	.01	6.38
	3	4.39	.60	.10	1.50	–	.01	6.60
	4	4.68	.60	.10	1.50	–	.01	6.89
	5	4.85	.60	.10	1.50	–	.01	7.06
Fossil-Fired Land-Based	5	4.16	.30	–	3.50	.30		8.71
Fossil-Fired BMPS	2	3.51	.60	–	2.75			6.86
	3	3.69	.60	–	2.75			7.04
	4	3.87	.60	–	2.75			7.22
	5	4.07	.60	–	2.75			7.42

Table 2.5   Conservative Estimate of Impact of the BMPS Concept on the Cost of Electricity (mills/Kwhr for a 1,000 MWE plant (1970)).

The next three rows of the table show the savings for a very interesting concept, namely, the fossil-fueled BMPS. Here we assume that a fossil station moored approximately three miles offshore can realize two important cost reductions: super-tankers can be used to deliver oil directly to the plants (impossible for onshore plants due to shallow water); and high-sulfur fuel might be burned if the atmospheric dispersion factor is high. The maximum savings of 1.67 mills/Kwhr is now large enough to help eradicate much of the chronic high power cost imposed on

Plant Design (1,000 MWE)	1970 Cost of Electricity at Busbar (mills/Kwhr)
Land-Based Nuclear (5-year construction)	8.05
Barge-Mounted Nuclear (4-year construction)	6.89
Barge-Mounted Nuclear (3-year construction)	6.60
Land-Based Fossil (5-year construction)	8.71
Barge-Mounted Fossil (4-year construction)	7.22
Barge-Mounted Fossil (3-year construction)	7.04
ADVANCED PLANT DESIGN (2,500 MWE)	1995 Cost
Land-Based LWR (5-year construction)	6.69
Barge-Mounted LWR (5-year construction)	5.79
Land-Based LMFBR (5-year construction)	6.50
Barge-Mounted LMFBR (5-year construction)	5.60

Table 2.6 Economic Comparison of Alternatives

New England's industrial customers, and a step in the right direction for residential power costs.

Finally, the last four rows project the study to 1995 when plant sizes should be approximately 2,500 MWE and when the liquid-metal-cooled fast-breeder (LMFBR) should supersede the light-water reactors (LWR). As can be seen, the combination of the LMFBR and BMPS could deliver busbar power approximately 3 mills/Kwhr cheaper than present-day fossil-fueled stations, even without the potential savings due to shortened construction time.

The results reported in Table 2.6 substantiate two important conclusions that may be drawn from this study. First, shorter construction times and lower pollution-abatement costs represent

sources of the largest potential savings for the BMPS concept;
and second, advanced nuclear concepts may provide an even further
reduction in electric power costs in the foreseeable future.

3. Other Economic Considerations

There are, of course, many implicit assumptions buried in
the analyses leading up to the figures presented in the preceding
section. Some of the more important ones will be elaborated
upon here in order to cast the results in the proper perspective.

First of all, it was assumed that land-based and barge-
mounted stations had equal construction costs of some 200 million
dollars. This assumption was based on the cost breakdown of
Table 2.3 which identified major cost differences between the
the two concepts. The BMPS was actually found to be seven million dollars cheaper than the shore station in the category of
Structures and Improvements. On the other hand, the BMPS cost
an extra three million dollars for underwater power transmission
cable. Because of the small net difference which was within the
estimated uncertainty of the analysis, both stations were therefore assumed to cost the same--200 million dollars.

Other small differences exist which were glossed over in
the preceding discussion. The increased productivity of shipyard labor over site labor might save as much as 0.1 mill/Kwhr in
ultimate costs of electricity; on the other hand, it will cost
more to operate and maintain the less accessible BMPS. These
various trade-offs result in an apparent stand-off insofar as a
net cost difference is concerned, and we are left with the compression in schedule as the major factor resulting in a net
positive cost differential.

In the cost comparison no allowance was made for the small
difference between the costs of salt water vs. fresh water condenser tubing. Thus in Table 2.6 the land-based costs may
also be equated to onshore ocean-site costs. The cost of the
barge was estimated to be 19 million dollars representing construction costs in a U.S. shipyard. If foreign construction is

permitted, a five million dollar savings can be realized on this
item.

Again we should emphasize that the entire cost comparison
is based upon the cost of electricity at the busbar of the plant.
Edison Electric Institute statistics show that the residential
consumers in New England paid an average of 26.4 mills/Kwhr for
electricity in 1969 (see Table 2.1). Thus transmission and distribution costs are sufficiently large so that the economies discussed in this report will not appreciably decrease residential
power costs. Large industrial users, however, can select site
locations and schedule use of off-peak power to bring their costs
down to near the 7 mill/Kwhr busbar price quoted for new plants.
Extremely large users such as industrial complexes could possibly
own or contract for their own power station to reduce costs further. Thus, the economic impact of the BMPS concept would be felt
indirectly through product cost reductions which might improve the
competitive position of regional industries.

4. Concluding Remarks

It should be emphasized that this economic analysis is very
preliminary in nature, intended to provide only a rough indication of the economic viability of the BMPS concept. In a detailed cost analysis the numbers will be very much a function
of the trade-off variables (associated with alternative offshore
designs) listed in Section IV such as seismic protection, distance from shore (pollution and transmission costs will vary),
environmental protection, etc. In such an analysis, the major
differential cost components would be as follows:

(1) Costs of alleviating thermal pollution.

(2) Interest costs and savings due to shipyard construction.

(3) Cost of real estate for plant site.

(4) Cost of transmission to shore.

(5) Shipyard construction costs.

(6) Cost of site preparation.

(7) Transportation costs.

(8) Insurance costs.

(9)   Cost of operations.

(10) Costs of environmental protection.

We have demonstrated the significance of potential savings
with the BMPS concept in the area of pollution control and construction costs. Results of the more detailed follow-up study
previously mentioned (see Reference 54) indicate that at least
one offshore BMPS design can be shown to be economically competitive (now) with a land-based facility. This is an encouraging
result since the analyses of that study and the one presented
here were purposely conservative in nature. For example, there
seems little doubt that as land prices continue to spiral upward (especially at the coast (see Chapter 3)), this component
alone will make the offshore station the most economically
feasible choice for the power companies. Other factors favorable to the BMPS concept that were not included are potential
savings associated with (1) closer location to load centers,
(2) better seismic protection, and (3) the development of a
learning curve in the production of large-scale installations.
All of these considerations could make sizable contributions to
savings over land-based plants in the near future.

We can conclude from the foregoing analysis that *the BMPS
concept has a high degree of economic feasibility and will become increasingly attractive in the near future* as issues unfavorable to land-based siting will undoubtedly continue in
presently-established patterns!

VI. SELECTED LEGAL AND POLITICAL ISSUES

Having concluded from our analysis that the offshore concept has a high degree of technological and economic feasibility, it is now important to consider the procedural framework
within which this concept might be realized. The purpose of
this section, then, is to provide an overview of the institutional environment in New England that is germane to the issues
of power-plant siting.[64] We shall consider two primary areas
of interest:

(1)   Legal constraints on power-plant siting;

(2)   Effects of political considerations.

1.   Legal Constraints on Power-Plant Siting

In this section, we examine the prevailing concepts in legal constraints on power-plant siting in New England and the present trends that have evolved in the wake of aroused public concern over the problems of environmental pollution.

(1)   Jurisdiction

Ocean water that is contiguous to the coastline out to *three miles* is considered to be state coastal water and is fully controlled by the contiguous state. Coastal land between the high- and low-tide marks is also under the jurisdiction of the state. Beyond the three-mile limit, the Federal Government has jurisdiction over fishing rights out to the territorial limit of 12 miles, while federal jurisdiction over the continental shelf (as determined by the Geneva Convention on the Continental Shelf) extends to a water depth of 200 meters. Since the technology and economics of offshore siting seem to indicate that the most feasible locations would be about a mile or so from shore, it is most likely that offshore stations would come primarily under the jurisdiction of state governments. Hence, the following legal concepts pertain to state authority in this connection.

(2)   Need for Building a Power Plant

A number of states require the utility to provide a determination of actual need for the construction of a power plant. For example:

Vermont[65]--"...No company...may begin site preparation or construction of an electric generating facility within the state...without the public service board having first found the same to promote the general good of the state..."

New York[66]--the Public Service Law stipulates that utilities must first obtain the approval of the Public Service Commission before beginning construction of gas or electric plants. The standards are public necessity and convenience, and engineering and economic feasibility.

Presumably, such requirements would be the same for a plant located offshore within the state's jurisdiction.

(3) Financing Approval

All New England states through their public utility commissions can approve or disapprove a utility's long-term (greater than one year) financial arrangements, i.e., borrowing or the issuance of stocks or bonds. Typical examples are the statutes in Massachusetts[67] and Maine.[68]

(4) Transmission Lines

Increasing concern has been voiced recently by local political action groups over the visual damage to the environment caused by the proliferation of overhead transmission lines. Overhead lines connecting offshore stations to inland loads would certainly be subject to such criticism. The alternative of underwater transmission would avoid conflicts of this sort. Many states have legal controls over the construction of trans- mission lines. For example:

New York[69]--applicants applying for a certificate of necessity for plant and transmission line construction must show that they have "received the required consent of the proper municipal authorities."

--the local authorities may require underground lines, but may not *exclude* all transmission lines.

While the construction of transmission lines usually comes under local zoning authority, the states have generally given the public utility commission "either the power to review municipal zoning decisions or exclusive jurisdiction over power- related land uses which preempts the authority of the local government to control such land uses."[70]

(5) Permits for Dredging, Filling, or Construction in Navigable Waters[71]

Several of the New England states regulate dredging or the placing of fill or support structures for transmission lines, etc., in their navigable waters. Concurrent jurisdiction over such activities is held by the U.S. Army Corps of Engineers.

The language of the Connecticut statute in this regard is typical:[72]

> ...No person...shall erect any structure, place any encroachment or carry out any dredging or other work incidental thereto in the tidal, coastal, or navigable waters of the state until such person...has submitted an application and has secured from the (water resources) commission a certificate or permit for such work and has agreed to carry out any conditions necessary to the implementation of such certificate or permit.

The States of New York,[73] Rhode Island,[74] Massachusetts,[75] Vermont[76] all have similar legislation. Also, Title 33 of the Code of Federal Regulations (CFR) states that structures across navigable waters must not hinder ships but gives no specific design criteria.

(6) Physical Occupation of Navigable Airspace

"Power-generating stations that include high structures such as smokestacks, radiation containment domes, and transmission towers may need to secure building permission from the Federal Aviation Administration (FAA) and/or the state aviation commission. The FAA jurisdiction essentially covers any construction more than 200 feet above ground level and any lower constructions that would enter an airport approach zone."[77] The license usually requires the installation of pertinent safety and warning devices. State aviation agencies may also review construction near airports in all New England states.

(7) Permits for Discharge of Cooling Water

The major legal constraints affecting the siting of (nuclear) power plants involves the effects of heated effluent on the local ecosystem. These constraints have been discussed in a recent report by the New England River Basins Commission.[78]

> Water pollution control laws in each state require that a permit from the state water pollution control agency be obtained before any matter may be discharged into the waters of that state. (Conn. Gen. St. § 25-54i(a); 38 Me. Rev. St. § 413; Mass. G.L. c. 21, § 43; N.H.R.S.A. § 149.8 (III); N.Y. Pub. Health Law § 1230, R.I.G.L. 46-12-4(b), 10 V.S.A. § 909.)

The controls required under the permit are based on
the classified standard and criteria for the quality of
the receiving water plus general objectives for the en-
hancement of water quality, protection of health and welfare, and considerations of present and future waste discharges. The permit applicant bears the burden of providing evidence that his discharge will be consonant with
the quality standard.

The Secretary of Interior has approved (with exceptions)
the water quality standards of the six New England States
and the State of New York. Included in the exceptions
were the criteria for temperature and the so-called "anti-
degradation" provisions. Modifications of these elements
have been completed or are being made by the state water
pollution control agencies.

On interstate waters, discharging matter which reduces the
water quality below a standard developed by the state with
review by and approval of the Secretary of the Interior
is subject to enforcement proceedings by the U.S. Attorney
General (33 U.S.C. § 466(g)(1)). Although not required by
federal statute, many state agencies have on their own
initiative developed quality standards for intrastate
waters.

Aspects of the federal/state standards which bear on power
plant siting decisions include the requirement that high
quality waters must be protected from degradation; there
is also the need to define temperature requirements for
maintaining an ecologically sound aquatic environment
and temperature requirements for mixing zones in which
the standards may not be applicable.

Site selection for thermal power plants is greatly affected
by water quality considerations since the facilities generally require large amounts of cooling water (about 1 million gallons of water per day per megawatt of plant capacity). The temperature of the discharge water is usually
about 20°F higher than the intake water. The physical
and biological effects on the receiving water of the discharged cooling water plus those of any chemical additives
can only be grossly estimated. Permits may be issued,
however, with provisions for corrective action as a result
of damages incurred during initial phases of operation.
As the technological base is expanded, more definitive
temperature requirements and perhaps systems of thermal
control will evolve.

The temperature criteria adopted by the states are quite
general as applied to coastal and marine waters. Massachusetts, for example, allows no temperature increase "except
where the increase will not exceed the recommended limits
on the most sensitive water use" (from classes SA, SB, SC,

Water Quality Standards, Comm. of Massachusetts, Water
Resources Commission, Division of Water Pollution Control,
adopted March 3, 1967). The most sensitive use is usually
the culture and propagation of shellfish. The limits are
recommended by the Divisions of Marine Fisheries, and for
similar fresh water standards, by the Division of Fish and
Game.

New Hampshire's Water Supply and Pollution Control Commission has been authorized to adopt the temperature criteria
and recommendations of the state fish and game department,
the New England Interstate Water Pollution Control Commission, or the National Technical Advisory Committee of the
U.S. Department of the Interior, selecting whichever set
provides the "most effective level" of control. (N.H.R.S.A.
§ 149:3 supp., para. V-a (1969 Acts, c. 337.) Similar
requirements in the other states create an important advisory role for the fisheries agencies in thermal discharge
permit proceedings throughout the region. Coordination
between water quality agencies and fisheries agencies
appears to be informal but effective.

An exception to the generality of most temperature criteria is New York's Criteria Governing Thermal Discharges,
approved in August 1969 (6 NYCRR 704.1). In most cases,
New York's rules provide clear guidelines for plant siting
and design decisions. The temperature of coastal waters,
for example, "shall not be raised more than 4°F over the
monthly means of maximum daily temperatures from October
through June nor more than 1.5°F from July through September except that within a radius of 300 feet or equivalent
area from the point of discharge this temperature may be
exceeded" (6 NYCRR 704.1).

The states all require that the permit be obtained before
the discharge commences. Common practice has been to construct a power plant, then to apply for a discharge permit before commencing to operate the plant. In this context, the discharge permit cannot operate as a formal factor in plant site selections. To be sure, the existence
and public knowledge of quality standards for receiving
waters can and do serve as an element in a power producer's siting decision. But the effects of a thermal discharge on the recognized uses of a water body--particu-
larly the effects on fisheries--remain very difficult to
predict or assess. Extensive biological studies, which
may include one or more years of investigation before the
discharge begins, may be needed in order to make sound
decisions about cooling water disposal.

The feeling that state water quality authorities should be
involved in such evaluations at very early stages of plant
site development has led to several recent amendments to
controlling legislation. The 1970 amendments to the

Federal Water Pollution Control Act include a new section requiring an applicant for a federal license or permit to file a certification from the relevant state water
quality standards control agency.

New York anticipated this procedure--in part--when the 1969
legislature amended the Public Health law to require that a
permit for a thermal discharge be secured before any person begins constructing a nuclear power plant (40 N.Y.
Pub. Health Law § 1140 (1969 Acts #86).

In addition, when such a party files a Preliminary Safety
Analysis with the U.S. Atomic Energy Commission for a
construction permit, he must also submit an environmental
feasibility report with the state Department of Health.
The New Hampshire Water Supply and Pollution Control Commission requires filing of plans for waste disposal devices, which will need discharge permits, at least thirty
days before construction begins.

(8)   Recent Trends in Legislation

        There have been two major trends in recent legislation
(evidenced in some of the previous sections) that have evolved
as a result of increasing concern over the effects of power-
plant siting on the environment. These are:

    (a) Requirement of a permit by the state which
         licenses *all* phases of a proposed power plant
         before any construction can begin (including
         obtainment of a preconstruction discharge per-
         mit before the construction of a *nuclear* power
         plant). Laws to this effect now exist in New
         York and New Hampshire; and

    (b) Federal law introduced to Congress in 1968
         giving the AEC authority in thermal pollution
         regulation for nuclear power plants.

    2. Extension of Legal Aspects to Offshore Siting

    We have already noted that, since the technology and the
economics of offshore siting indicate that the most feasible
location would be about a mile or so from shore, it is most
likely that offshore stations would come primarily under the
jurisdiction of state governments. Due to recent concern over

the effects of thermal discharges, legal mechanisms have evolved
to regulate the effects of coastal power plants on the water environment. There seems to be no reason why these mechanisms
would not apply directly to offshore plant siting. The only
area where regulations are not well developed (since the need
has not come up) concerns the effects of heated discharge on sea
life at a considerable distance from shore (1-3 miles). However, such questions seem to be easily handled within the framework of existing legislation. Due to the increased distance from
shore of an offshore station and the various methods for discharging heated waters in deep water in an ecologically-safe manner, it appears that offshore power stations would minimize the
ecological stress of thermal pollution and comply easily with
present and projected standards. Hence, there seem to be no formidable legal barriers to the realization of the offshore concept.

  3. Political Considerations Affecting Power-Plant Siting

  The primary political issues affecting the siting of offshore power plants involve public concern over the safety, environmental, and national security aspects of an ocean-based
power plant. We have already discussed the first two of these
within the legal framework, and assume that they will *not* be of
major consequence *if* the proper environmental and safety constraints are built into effective legal regulations governing
offshore siting. This leaves us with the question of national
security.

  Due to a general unfamiliarity with the criteria that might
be applied in this area, we can comment on this question only
in a very approximate way. It would appear offhand that an offshore power station might be more susceptible to attack from
conventional craft, particularly submarines, than its land-based
counterpart. Presumably, the dangers from attack by conventional
aircraft and nuclear weapons would be about the same, although
there might be some advantage to having the power station some
distance out to sea in the event of nuclear warfare (isolation
from ground disturbances). The important variables (as they

were with the technology and the economics) seem to be distance
from shore and depth of water. If a floating station were to
be located one mile offshore in 50 feet of water and surrounded
by a breakwater, it seems likely that adequate protection could
be provided using World War II-type submarine nets, electronic
surveillance devices, and the natural barrier provided by the
breakwater. Since this appears to be the type of alternative
that is most economically and technically feasible, this aspect
of the question of national security would appear resolvable.

We do not feel adequately equipped to discuss the pertinent
strategies of defense in any greater detail, and leave a further
examination of national security issues relating to the offshore
concept to the higher councils of government.

VII. CONCLUSION

The preceding sections of this chapter have explained in
some detail the rationale behind selection of the barge-mounted
power station concept, and have presented a technical and economic evaluation of the proposal. To recapitulate just the key
characteristics of the BMPS: it offers a way to avoid serious
conflicts over the use of scarce land resources; it has the
potential to reduce electric power costs to industry in the New
England Coastal Area to a level competitive with the national
average, while at the same time eliminating inland and shoreline
air and water pollution problems associated with power production; and the BMPS can also become a new regional export product
and serve to allocate in a more efficient way the resources of
the New England shipbuilding industry. The five major conclusions are:

(1) *The BMPS is technically feasible with only minor ex-
tensions of current engineering capabilities (and is adaptable
to nuclear, fossil fuel, or even magnetohydrodynamic (MHD)
generating plants);*

(2) *It could eliminate completely the harmful contributions of the electric power industry to environmental pollution;*

(3) *It drastically reduces construction times to help meet increasing demands and reduce costs;*

(4) *It avoids serious conflicts in land use;*

(5) *It is economically attractive and legally plausible.*

This last attribute is particularly interesting since it should facilitate implementation of the idea without the need for devising new ways to see that the private market adequately weighs the public interest. In brief, there appears to be no impediment to realization of this concept that cannot be resolved through due process using the existing political and economic system.

What now are the implications of this analysis of the off-shore concept? It is clear that this concept provides a viable alternative to land-based power stations and can eliminate some of the most serious problems faced by the power industry today in meeting our nation's increasing demands for electric power. All indications are that the concept is feasible from a wide range of technological, political, legal, and economic stand-points. What then can be done to set the machinery in motion to give the implementation of this concept a long hard look?

To answer this we must first decide what machinery we are talking about. In our country today, there is no centralized governmental body at the federal level that is charged with the formulation of national goals and priorities and the long-range planning that is necessary if we are to meet our future needs for electric energy. Yet it would be difficult to imagine a more pervasive issue in relation to the maintenance of our society as it exists today. The aspirations of every American citizen for a greater level of well-being are based in part upon the confidence that this nation can maintain the capability to provide adequate supplies of electric power on a continuing basis. Yet today we live under the recurring threat of brownouts and blackouts in our major cities each summer, while evidence of any

substantive effort at the federal level to attack the root causes
of these problems remains conspicuously lacking. This is not to
say that there is a lack of concern--certainly the Atomic Energy
Commission has been instrumental in developing the technology
that we need to meet increasing demands. But a number of the
problems of power production, as we have seen (e.g., plant
siting), are not technological in nature; rather, they are social
problems generated by an awakened sense on the part of our society of the value of the environment and of the mistakes we have
made for so long in pursuing a course of unbridled growth. Yet
there is no existing mechanism to direct our course and help resolve conflicts in a manner consistent with a carefully considered set of national objectives and policies in the area of
electric energy production. Hence, the path of power generation continues to wander helter-skelter in whatever direction
is randomly dictated by the combined activities of the private
marketplace and local political decision-making. Is it wise to
entrust such a crucial issue as electric power production to
anyone other than the highest level of government where the true
national interest can be fully determined and appreciated? We
think not! It must be emphasized that this is not a matter to
be resolved at the state or local level--the problems of air and
thermal pollution and land use (especially of coastal resources)
in relation to power production are regional and national in
scope. Nor can it be the responsibility of the electric utility
industry unless present forms of regulation are modified to
allow for the generation of the large amounts of capital that
would be required to fund a large-scale research and development
effort. The responsibility lies clearly at the federal level
where the power production issues related to land-use management
and a host of other areas of national concern can only be resolved as part of coordinated effort at the formulation of a
national energy policy!

The implication is clear--we are in need of a comprehensive,
long-range, coordinated effort at the federal level that will
bring together all those areas of concern that are affected by

the lack of a national energy policy. Certainly the problems
of land-use management, environmental pollution, and adequate
provision of power to meet necessary demands are foremost among
these areas of national concern. It is precisely these problems
that have been attacked in this analysis. Hence, *serious consideration of the offshore concept is particularly germane to
the formulation of a national energy policy and should be effected through an in-depth study of the concept as a viable alternative to land-based installations, bringing together federal
agencies including the Office of Science and Technology, the
Atomic Energy Commission, the Department of the Interior, and
any public or private organization with a vested interest in
land use, environmental quality, and other issues related to
power-plant siting.* Such an effort could be instrumental in
eliminating a number of troublesome sources of controversy; at
the same time, this could be the all-important first step that
would draw attention to the increasing need for the establishment
of a formal governmental mechanism that can effectively deal with
the formulation of a long-range, national energy policy!

APPENDIX A

The unit cost of electricity for a given power plant is given by:

$$e = \frac{\Phi}{8.76L} \frac{I}{K} \quad + \quad F \quad + \quad 0$$
$$\text{(fixed charge} \qquad \text{(fuel cost} \qquad \text{(operating cost}$$
$$\text{component)} \qquad \text{component)} \qquad \text{component)}$$

where

      e = unit cost of electricity (mills/Kwhr)

      Φ = annual fixed charge rate

      I = initial cost of plant (dollars)

      K = rated net capacity (Kwhr)

      L = load factor (percent)

      F = fuel cost component (mills/Kwhr)

      0 = operating cost component (mills/Kwhr)

Using this equation and the ground rules presented in Table IV, electricity costs (in mills/Kwhr) can be calculated for new fossil- or nuclear-fired power plants.

	Nuclear		Fossil	
	U.S. Average	New England	U.S. Average	New England
Capital Charges	4.95	4.95	4.16	4.16
Operation and Maintenance	.30	.33	.30	.33
Fuel	1.50	1.50	2.70	3.50
Nuclear Insurance	.10	.10	-	-
Total	6.85	6.88	7.16	7.99

Source: Manson Benedict, "Economics of Nuclear Power," notes associated with Course 22.27 given at the Massachusetts Institute of Technology, Department of Nuclear Engineering (Spring 1970).

REFERENCES

1. W. G. Jensen, Nuclear Power, G. T. Fouks & Co., Oxfordshire, England (1969), p. 41.

2. Edison Electric Institute, Statistical Yearbook of the Electric Utility Industry for 1966, New York (1967), p. 50.

3. Ibid., p. 50.

4. See Reference 1, p. 141.

5. Ibid., p. 143.

6. Ibid., p. 142.

7. See statements of Philip Sporn in a letter to Senator John A. Pastore, Chairman of the Joint Committee on Atomic Energy, (J.C.A.E.), from Nuclear Power Economics - 1962 through 1967, Report of the J.C.A.E. of the United States (1968), pp. 2-20.

8. H. G. Lawson, "Nuclear Split," Wall Street Journal, May 20, 1970, p. 1.

9. S. Novick, The Caveless Atom, Houghton Mifflin Co., Boston, Mass. (1969).

10. R. Curtis and E. Hogan, Perils of the Peaceful Atom, Doubleday, Inc., Garden City, New York (1969).

11. D. Farney, "Atom-Age Trash - Finding Places to Put Nuclear Waste Proves a Frightful Problem," Wall Street Journal, January 25, 1971.

12. Sherman R. Knapp, Chairman of Northeast Utilities; Statements in Panel Session on "The Nuclear Controversy," Annual Conference, Atomic Industrial Forum, December 1969.

13. The Chase Manhattan Bank, Energy Division, Outlook for Energy in the United States, New York, October 1968, p. 47.

14. Gordon R. Corey, economist for Commonwealth Edison (Illinois); from nuclear engineering lectures given at M.I.T., April 7, 1970, entitled "Effects of Recent Changes in the Money Market."

15. Estimate based on information assembled in connection with updating of the 1964 Federal Power Survey; reported in Considerations Affecting Steam Power Plant Selection, a report sponsored by the Energy Policy Staff, Office of Science and Technology, December 1968.

16. U.S. Department of Health Education and Welfare, The
     Sources of Air Pollution and Their Control, 1966.

17. Report of the Energy Policy Staff, Office of Science and
     Technology, Considerations Affecting Steam Power Plant
     Selection, USAEC Report TID-24936, December 1968.

18. Glenn T. Seaborg, Chairman of the Atomic Energy Commission,
     testimony before J.C.A.E. hearings on "The Environmental
     Effects of Producing Electric Power."

19. F. C. Olds, Nuclear Plants - Late and Later, Power Engi-
     neering, March 1969.

20. See Reference 17.

21. Senator Henry M. Jackson, "Introduction of the National
     Land Use Policy of 1970," Congressional Record-Senate,
     January 29, 1970.

22. Joint Committee on Atomic Energy, Environmental Effects of
     Producing Power, August, 1969.

23. Ibid., Part 2, January-February, 1970.

24. Harris B. Stewart, Jr., "The Turkey Point Case, Power Devel-
     opment in South Florida - a Study in Frustration,"
     Meeting of the Marine Technology Society, November 1970.

25. J. A. Carver, Jr., statements in panel session on "The
     Nuclear Controversy," Annual Conference, Atomic Industrial
     Forum, December 1969.

26. From survey made for the Federal Power Commission by its
     Northeast Advisory Commission, Nuclear Industry, April
     1969, pp. 27-28.

27. Edison Electric Institute, Statistical Yearbook of the
     Electric Utility Industry for 1969, New York (1970).

28. See Reference 22.

29. General Dynamics, Quincy (Mass.) Division, pamphlet,
     "Proposed Fast Containerships as Part of a Total Trans-
     portation System" (1968).

30. Based on data presented by R. W. Marble (Reference 20) and
     projected demands for electric energy consumption.

31. "Con Ed Has Taken the First Step Toward a Serious Look at
     Artificial Islands," Nucleonics Week, Vol. 11, No. 14,
     April 1970.

32. Betchel Corporation, Engineering and Economic Feasibility
    Study Phases I, II, III for a Combination Nuclear Power
    and Desalting Plant, USAEC Report TID-22330, Vols. 1 and
    2, December 1965.

33. News items from Nuclear Engineering International, Vol. 15,
    No. 169, June 1970, and Ocean Industry, Vol. 4, No. 12,
    December 1969.

34. "Undersea Sites Projected for Nuclear Power Plants," New
    York Times, August 19, 1970.

35. R. W. Marble, "A Submerged Commercial Power Generating
    Concept," General Dynamics, Quincy Division, January 1970
    (unpublished).

36. R. M. Bunker, "Description, Application, and Siting of the
    STURGIS (MH - 1A) Floating Nuclear Power Plant," Trans.
    Amer. Nuclear Society, Vol. 8, No. 1, June 1965.

37. Staff of First Atomic Ship Transport, Inc., "Operating
    Experience of the N.S. Savannah in Commercial Service,"
    Nuclear Safety, Vol. 8, No. 1, Fall 1966.

38. W. Wiebe, et al., "Safety-Design Aspects of Nuclear Powered
    Ship Otto Hahn," Nuclear Safety, Vol. 9, No. 6, November-
    December 1968.

39. T. D. Anderson, "Offshore Siting of Nuclear Energy
    Stations," Oak Ridge National Laboratory, 1970
    (unpublished).

40. Daniel, Mann, Johnson and Mendenhall, "A Floating Earth-
    quake-Resistant Nuclear Power Station," USAEC Report
    TID-24558, April 1968.

41. H. G. Arnold, W. R. Gall, and G. Morris, "Feasibility of
    Offshore Dual Purpose Nuclear Power and Desalination
    Plants," USAEC Report ORNL-TM-1329, January 1966.

42. H. M. Busey, "Floating Plants for Seismic Protection,"
    Nuclear Applications, Vol. 6, June 1969.

43. See Reference 40.

44. See Reference 35.

45. O. H. Klepper and C. G. Bell, "Underwater Caisson Contain-
    ment of Large Power Reactors," USAEC Report ORNL-4073,
    Oak Ridge National Laboratory, June 1967.

46. See Reference 36.

47.   Edison Electric Institute, "Fast Breeder Reactor Report,"
      New York, April 1968.

48.   See Reference 35.

49.   Professor Ernst Frankel

50.   J. S. Watts, Jr., and R. E. Faulkner, "Designing a Drilling
      Rig for Severe Seas," Ocean Industry, November 1968,
      pp. 28-37. See also Graham, "Mooring Techniques in the
      Open Sea," Marine Technology, Vol. 2, No. 2, April 1965.

51.   Ocean Industry, Vol. 4, No. 12, December 1969, p. 28.

52.   Ocean Industry, Vol. 5, No. 6, June 1970, p. 64.

53.   Combustion Engineering Corp., "Liquid-Metal Fast Breeder
      Reactor Design Study," CEND-200 (1964).

54.   "Offshore Siting of Nuclear Power Generating Units," results
      of investigations of the technological feasibility and
      economic viability of the offshore concept in Course
      22.33, given at M.I.T. in the Nuclear Engineering Depart-
      ment, Fall 1970, internal document MIT NE-120.

55.   Ibid.

56.   See Reference 42.

57.   See Reference 54.

58.   J. J. Dinunno, F. D. Anderson, K. E. Baker, and R. C. Water-
      field, "Calculation of Distance Factors for Power and
      Test Reactor Sites," TID 14844, Division of Licensing and
      Regulation, U.S.A.F.C. (1962).

59.   See Reference 41.

60.   See Reference 40.

61.   See Reference 42.

62.   W. Steigelmann, "The Outlook for Nuclear Power-Station
      Capital Costs," Reactor Technology, Vol. 13, No. 1,
      Winter 1969-1970.

63.   Ibid.

64. This treatment is based on a recent report by the New
     England River Basins Commission (NERBC), Laws and Proce-
     dures of Power Plant Siting in New England, Power and
     the Environment/Report No. 1, February 1970.

65. Vermont Stat. Ann. § 246.

66. N.Y. Public Service Law § 68.

67. Massachusetts General Law, Chapter 164, Sec. 14.

68. Maine 35 Rev. Stats. s. 172.

69. See Reference 66.

70. See Reference 64, p. 11.

71. See Reference 64, pp. 23-24.

72. Connecticut General Statutes 25-7d.

73. N.Y. Conservation Law, § 429-b.

74. Rhode Island General Laws, 46-6-2.

75. Massachusetts General Law, Chap. 91 § 14.

76. 29 Vermont Stat. Ann., § 553.

77. 14 Code of Federal Regulations, Part 77, Objects Affecting
     Navigable Air Space.

78. See Reference 64, pp. 32-33.

79. See Reference 64, pp. 19-23.

CHAPTER 3

THE CRISIS IN SHORELINE RECREATION

by

Dennis W. Ducsik

Contributing Author: Robyn Seitz

ABSTRACT

Our nation today faces a crisis in shoreline recreation. It has come about because a mushrooming demand for the unique and rela- tively scarce resources of the coastal zone has far outstripped the available supply. We have allowed a pattern of economic growth and development in the coastal zone to continue unchecked for the past three hundred years, so that now we find that only a small percen- tage of the entire shoreline is in public hands for recreation. The problems of pollution and erosion have combined with the in- creasing tendency of private owners to restrict public access so that the supply of available shoreline, limited to begin with, is shrinking steadily. Yet the demands are increasing at a breakneck pace. The multiplicative effects of increasing population, income, leisure time, and mobility are expected to bring about a tripling in the demand for outdoor recreation by the turn of the century. Yet the facilities are saturated *today* with hordes of users, while there is little or no room for expansion within the existing econo- mic and political environment.

This serious problem has materialized because of imperfections in our present allocative mechanisms of the private market and local political decision-making. Analysis has shown how these mechanisms fail to provide an efficient allocation of valuable re- sources in particular circumstances. These circumstances include (1) the inability of the price system to determine and articulate the true costs and benefits to society associated with a particu- lar good and (2) the tendency of local political bodies to make de- cisions based on effects that are net benefits to the local commu- nity but not to the regional society.

A new framework for long-term coastal zone management is pro- posed that places the prime responsibility for shoreline regulation in the hands of the States. At the same time, it must be recog- nized that a strong federal involvement is necessary to (1) coor- dinate the efforts of individual States and resolve conflicts that arise due to interstate secondary benefits and (2) establish uni- form objectives and guidelines to assist the States in the problem of *how* decisions are to be made in the absence of the private market discipline.

Recognizing the need for action in the short- as well as the long-run, we have focused on two areas that are subject to heavy demands for shoreline recreational opportunities: Cape Cod and the Boston Metropolitan region. We propose that the South Cape Beach and the Boston Harbor islands can and should undergo devel- opment for recreational purposes in the very near future.

90

CHAPTER 3

THE CRISIS IN SHORELINE RECREATION

I. INTRODUCTION

Over the last three hundred years, the American shoreline
has been considered a plentiful resource to be used freely by
man for the growth and progress of his society. Since the times
of the early colonists, the coastal areas have been the gateways
to this nation. The first settlements that grew up around the
natural harbors of the coastal zone have since developed into
thriving centers of population and industry and are now focal
points for the transportation and commerce of our nation.
Throughout this historical period of population growth and industrial expansion, the coastal zone has been recognized as an
attractive place to live and work, a convenient transportation
corridor linking the coastal cities, and an ideal source of
recreational opportunity. Since the capacity of coastal resources
to support these multiple endeavors has always been viewed as
adequate, "the laws regulating man's activities in this zone
were historically intended to protect and serve individual and
group interests in dealing with each other"[1] within the context
of the economic system of free enterprise in the private marketplace.
Under this system, the shoreline "has largely been left for
acquisition and exploitation by whatever public or private agencies
desired to undertake its ownership, control, and management."[2]
Since there always seemed to be plenty of shoreline open for
a wide variety of recreational pursuits and no indications of
serious damage to ecological systems in the estuarine zones,
there was no perceived need for public interference in the allocative workings of the private market. The result is that, today,
approximately 91 percent of this limited, unique natural resource
is under private control, another 3 percent is restricted for
military uses, leaving only 6 percent of the shoreline in public
ownership. Thus, the coastline, as a public commodity, has
become one of the most scarce of all our valuable natural assets,

91

extremely short in supply relative to the heavy demand from
competing uses.

     Under normal market conditions, the prices associated with
coastal real estate would adjust in such a situation so that the
use deriving the greatest benefit (as measured in ability and
willingness to pay) from coastal ownership would be able to secure control. This is indeed happening to a certain extent as
the cost of acquiring shoreline property has become astronomical
in recent years. *It has become increasingly clear, however, that
the price mechanism of the private market has failed to represent
certain important societal values in its allocation of coastal
resources and is unable to provide for the proper expression of
those values in a competitive marketplace.* The two most often
misrepresented values, the first of which is the topic of discussion in this article, are those associated with (1) the provision
of adequate facilities for *outdoor recreation* in the coastal zone,
and (2) the protection and preservation of the *ecological systems*
that abound in the marine environment. Although both of these
are intricately related to the life and livelihood of every
person in modern society, the coastal areas of this nation have
been sorely neglected as a public resource, while the need for
careful allocation of this irreplaceable asset has gone unattended.
For generations there was little or no awareness of the dangers
to future society "as long as the ability of the natural environment
to absorb the effects of the socioeconomic environment seemed
unlimited, and the problems of pollution and environmental damage
were isolated."[3] Only recently has it become apparent that
"the laws protecting man from himself must be extended to protect
the natural environment from man."[4] We have begun to recognize
the need of human society for the resources of the coastal zone
and its value to civilization both as an essential part of its
ecosystem and as an exploitable asset to be carefully allocated
among competing uses. Of all these competing uses, the two
that are most often misrepresented--recreation and ecology--
may ultimately turn out to be the most important to the long-
term health and well-being of man in our modern society! Although

man is now the dominant species on earth, his very survival
depends on the intricately complex ecological balance among
all plants and animals within their respective environments;
and the well-being of each individual depends upon the maintenance
(with the help of recreation) of his internal psychological
balance and the external balance that exists in his interactions
with the outside world.   The need for recognition of this dual
value in our coastal resources has been emphasized in the National
Estuarine Pollution Study:[5]

> It is the value of the estuarine zone as a fish and
> wildlife habitat, a recreation resource, and an esthetic
> attraction that make [it] a unique feature of the human
> environment, yet it is these very values that have been
> generally ignored in satisfying the immediate social and
> economic needs of civilization.

All this points to the importance of the allocation of our
scarce, valuable shoreline resources as an issue in land-use
management.   It is the purpose of this chapter to set forth the
economic, political, and sociological aspects of coastal land use
for *outdoor recreation*, with a focus on the New England shoreline.
The goal is to derive some insight into the nature of effective
land-use policies that might be used to govern the allocation of
shoreline resources in a manner most consistent with the goals
and values of American society.

II. THE STATUS OF SHORELINE RESOURCES

1. Background

Our nation faces a crisis in shoreline recreation, right
now, today.   The mushrooming demand for this unique and relatively scarce resource has far outstripped the effective supply.
The problems are particularly acute in the crowded Northeast of
which the New England region is a part.   Anyone who has been delayed for hours on a hot day in bumper-to-bumper traffic to Cape
Cod beaches, who has experienced the mobs of people at the Revere
and Lynn shores, or who has not been able to get to the coast at
all because the beach was closed due to pollution or filled-to-
capacity parking facilities, will attest to the immediacy of

this critical shortage of available, accessible shoreline recreation areas.

All indications are that unless immediate action is taken,
these problems will get much, much worse. The demand for outdoor
recreation, especially at the shore, has increased significantly
in the last ten years. The trends toward more leisure time, more
real income, and greater mobility enable larger proportions
of our growing population to seek and enjoy recreation activity
of all types each year. The effects of these trends on outdoor
recreation are evidenced in part by the rapid growth of companies making equipment for use in outdoor activities, and in the
large increases in service facilities (such as campgrounds) that
support the recreationalist in his varied pursuits. This gives
us an indication of what to expect in the future: "more people
taking more vacations, learning more about vacations and recreation, developing a wider range of skills and making more demands
on every kind of recreation area, and rearing a generation of
outdoor-minded children who will have even more skills and make
even more demands."[6]

The critical nature of this situation is aptly described by
Bayard Webster, of the *New York Times*:[7]

> The shoreline of the United States has been so built up,
> industrialized and polluted during the last decade that
> there are relatively few beaches left for the family in
> search of a free, solitary hour by the sea.
>
> From Maine to Florida and on around to Texas, from South-
> ern California up to Washington State, the nation's sea-
> shores have become cluttered with hotels, motels, sprawling
> developments, military complexes and industries of every
> kind.
>
> Miles of tranquil beaches where hundreds of seaside re-
> treats were once open to everyone for swimming or fishing
> have been fouled by oil spills, industrial effluents,
> farm pesticides and city sewage.
>
> What remains--shoreland that is not dirty, crowded or closed
> to the public--amounts to a tiny fraction of the country's
> total coastal zone, about 1,200 miles or 5 percent of the
> shore areas considered suitable for recreation or human
> habitation.

The prospect of continuing encroachment, together with the intensified natural erosion often caused by heedless development (even in normal weather, winds and waves can eat away or shift up to 20 feet of beach a year), has alarmed many marine biologists and conservationists.

Although...conservationists have been encouraged by indications that some states and bureaus of the Federal Government are becoming interested in protecting the nation's coastline as a separate national resource, they fear that it may already be too late to reverse the trend.

Close to the heart of the problem are two factors largely beyond the control of governmental authorities....One is the sharp increase in recent years in the nation's population. The other is the rush to the large coastal cities by millions of people from inland rural areas.

The result is that popular demand for open recreational space near the water is rising just as private and industrial developers are fencing off the best of it--if not the last of it in any given area--and land prices are spiraling far beyond the means of most urban dwellers.

In this article, Mr. Webster has struck at the heart of the issue from all its crucial aspects. First, the dwindling supply of shoreline recreational areas has been caused in part by the *acquisition* of coastal acreage for use by large *industrial and commercial complexes*. Our historical emphasis on economic growth and industrial expansion has allowed this to happen without the full realization of the extent to which such uses exclude all others. As a result, 40 percent of all the manufacturing plants in the United States today are located within the borders of the coastal counties. This is clear evidence of the consequences of nonexistent land-use planning. The use of coastal acreage for industrial or commercial purposes may be necessary for some enterprises with a demonstrated need for ocean accessibility. For example, some industries (tank-oriented oil companies and chemical plants) require multi-fathom harbors, while others (paper, primary metals, power generation) require substantial water supplies in the course of normal operations. Yet use of coastal land for these uses alone has resulted in the loss of many miles of scenic shoreline. In California, for example, power companies have occupied large stretches of the coast for the siting of

power-generating facilities. Even for industries such as this,
ways should be sought to satisfy the operational requirements
for water while minimizing the usurpation of coastal land to
meet these needs (see Chapter 2).

Second, the expanding and coastward-shifting *population* has
placed fantastic pressures on the shoreline for private development. This trend is accelerated by continuing increases in
disposable income, leisure, and mobility. The demand for vacation homes and resort communities by the sea has sent land
values skyrocketing. In South Carolina, the price of a front-
foot of shoreline is $1,600, while in Massachusetts the price
of an acre of shoreland has increased by a factor of five since
1965 to $50,000. Even the relatively wild areas of North
Carolina and Maine, far removed from population centers and
lacking in good transportation facilities, are now in the hands
of speculators who are assured a fantastic profit in the not-too-
distant future. They are well aware of how the craving for vacation space by the ocean "has led to the development in such
places as Virginia Beach, Virginia, and Ocean City, Maryland, of
coastal sections in which houses, motels, and hotels are built
as close as six feet apart for many miles along the beach."[8]

A third major factor contributing to the decrease in available shoreline areas for recreation is *pollution*, which has destroyed countless fish and shellfish areas and fouled beaches in
and around every major coastal city. In Boston Harbor, many
islands would offer excellent opportunities for a variety of
water-related activities were it not for the poor water quality,
due in part to high bacteria counts resulting from municipal
sewage dumping and storm sewer overflow. Oil spills, pesticides,
and industrial effluents have also taken their toll of valuable
shoreline resources. The accelerated eutrophication of Lake Erie
is probably the most celebrated example of this serious problem.

A final element contributing to the decreasing supply of
coastal land is *shore erosion*, which is often accelerated by
improper land use that stems from an ignorance of the dynamics

of beach areas. A recent article entitled "America's Shoreline
is Shrinking"[9] points out the seriousness of this problem:

> From Cape Cod to California, America's ocean shoreline
> is being cut and furrowed by erosion. Much of this is
> the result of the ceaseless action of waves and wind, a
> combination of forces as old as the sea itself....[an
> example is] the dramatic case of Cape May, New Jersey,
> a famous resort area which has lost a fourth of its land
> area to the combined action of wind and wave during the
> last 30 years or so.
>
> The State of Maryland loses about 300 acres of valuable
> land every year along the shores of Chesapeake Bay....
> Sections of shoreline at Point Hueneme, California,...
> have receded as much as 700 feet in ten years.

The article goes on to point out how the natural forces of
erosion are greatly abetted by the actions of man. Joseph B.
Browder, a southern field representative for the Audubon Society,
has cited erosion in Miami Beach "caused by hotels built almost
right in the surf, housing projects built on thousands of once-
wild acres of tidal marshes."[10] Ian McHarg, in his book Design
with Nature,[11] has pointed out the dangers that trampling dune-
grasses, lowering the level of groundwater, and interrupting
littoral sand drift pose to the stability of dune formations.
He has this to say about such formations in New Jersey:

> The knowledge that the New Jersey Shore is not a certain
> land mass as is the Piedmont or Coastal Plain is of some
> importance. It is continually involved in a contest
> with the sea; its shape is dynamic. Its relative stabi-
> lity is dependent upon the anchoring vegetation....
> If you would have the dunes protect you, and the dunes
> are stabilized by grasses, and these cannot tolerate
> man, then survival and the public interest is well served
> by protecting the grasses. But in New Jersey they are
> totally unprotected. Indeed, nowhere along our entire
> eastern seaboard are they even recognized as valuable....
> Sadly, in New Jersey no...planning principles have been
> developed. While all the principles are familiar to bot-
> anists and ecologists, this has no effect whatsoever upon
> the form of development. Houses are built upon dunes,
> grasses destroyed, dunes breached for beach access and
> housing; groundwater is withdrawn with little control,
> areas are paved, bayshore is filled and urbanized. Ig-
> norance is compounded with anarchy and greed to make the
> raddled face of the Jersey Shore.

2. The General Picture

A summary of the tidal shoreline of the United States as reported by the U.S. Coast and Geodetic Survey (excluding Alaska and Hawaii) is given in Table 3.1. The shoreline is one of our most popular resources for outdoor recreation and is in heavy demand; yet, as the table shows, it is most scarce in terms of public ownership for recreation. The 48 contiguous states have nearly 60,000 miles of shoreline, of which about one-third is considered suitable for recreational activities. This possible recreation shoreline includes beach, bluff, and marsh areas that must meet the following criteria:[12]

(1) The existence of a marine climate and environment;

(2) The existence of an expanse of view of at least five miles over water to the horizon from somewhere on the shore;

(3) Location on some water boundary of the United States.

Shoreline Location	Detailed Shoreline Stat. Miles	Recreation Shoreline Stat. Miles	Public Recreation Stat. Miles	Restricted Stretches Stat. Miles
Atlantic Ocean	28,377	9,961	336	263
Gulf of Mexico	17,437	4,319	121	134
Pacific Ocean	7,863	3,175	296	127
Great Lakes	5,480	4,269	456	57
TOTAL	59,157	21,724	1,209	581

Mileage of detailed shoreline, recreation shoreline, public recreation shoreline, and restricted shoreline by major coastlines as measured using Coast and Geodetic Survey methods and meeting criteria defined in text.

Source: Outdoor Recreation Resources Review Commission Study Report No. 4 (1962), p. 11.

Table 3.1 Tidal Shoreline of the United States

        The figures presented in the table indicate that less than
two percent of the total shoreline is in public ownership for
recreation, while only about 5.5 percent of the recreational
shoreline is in public hands. On the entire Atlantic Coast,
only 336 miles of shoreline are publicly owned for recreation,
a mere three percent of the total recreational shoreline. Yet,
this coast contains the population concentration of the sprawling
Northeast megalopolis and Florida. Near these metropolitan
areas, the demands are the greatest, yet the available absolute
supply is small. Nationally, the coastal areas contain about
15 percent of the total land area, "but within this area is
concentrated 33 percent of the nation's population, with about
four-fifths of it living in primarily urban areas which form
about 10 percent of the total estuarine zone. The estuarine
zone then is nearly twice as densely populated as the rest of
the country."[13] Understandably, the disappearance of natural
beaches and other shoreline recreational resources is most evident
near these most populous areas. "From Massachusetts to North
Carolina, in Florida, in California near Los Angeles and San
Francisco and along the Gulf Coast, a sprawling confusion of
buildings crowd the shore."[14] When the added effects of pollution (most severe in metropolitan areas) and erosion on existing
supply are taken into account, the situation becomes even more
critical. In the densely-settled North Atlantic and Middle
Atlantic regions, there are 5,912 miles of recreational shoreline, of which 5,654 miles are under private or restricted public
control; hence, 97 percent of the shore is inaccessible to the
general public! Yet, the pressures on shoreline facilities
near metropolitan areas are so great that frequently the waters,
even in busy harbors, "are used for recreational purposes by
those who cannot afford to go elsewhere, regardless of whether
the waters are safe for body contact or not."[15] This points
to still another problem, the inability of low-income, less-
mobile groups to find suitable coastal recreational facilities
anywhere but in the immediate vicinity of urban centers, where
the pollution problems are most severe, and where fewer beaches

are available and oftentimes inaccessible due to gross overcrowding.

3. A Case Example

The critical magnitude of the supply situation with regard
to shoreline resources can best be demonstrated by considering
what has been happening in the State of Maine in recent years.
Maine's varied and beautiful shoreline is its greatest natural
asset. The coastal zone includes 10 percent of the total geographi-
cal area, 36 percent of the population, and 127 local governmental
units. Forty percent of the wages in Maine are generated in
this zone, while 60 percent of all recreational property and
seasonal residences are located there. Almost the entire coast
is steep, rocky bluff with occasional small beaches of gravel
or mud.[16] In many areas, deep water occurs close up to the
shore. The coast is very irregular with numerous coves, inlets,
small bays, and similar areas serving as harbors or sheltered
areas. The shore area is only slightly developed with only 34
miles (or 1.4 percent of the coastline) in public ownership for
recreation; the primary uses over the remaining 2,578 miles
are private with some commercial resort activity. The shoreline
is least suitable for swimming and water sports since there are
only 23 miles of beach along the entire coast. The most suitable
activities are camping, hiking, boating, sailing, and sightseeing,
for which the 2,520 miles of ragged, rocky bluff shore provide an
ideal setting. However, these activities are severely restricted
in many places due to extensive private ownership of prime coastal
property.

Pollution has caused some problems with the taking of shellfish. By 1962, 67,000 acres of tidal flats had been closed
to shellfishing, a source of income and enjoyment to residents
and visitors alike. In the decade preceding 1962, the total
areas closed due to pollution increased by 12 percent.

By far the most serious question facing Maine with regard
to its shoreline resources is the extremely small percentage
of public ownership. In 1967, a land-use symposium, organized
by land consultant John McKee, pinpointed the issues relating

to this question and outlined the successes and failures of
Maine's governmental bodies in dealing with it. McKee and his
colleagues emphasized the public's right of access to unique
shoreline, not only to a "mudflat or a rundown beach, but to
a cliff and forest and cove--precisely the places that are selling
fastest today....Unless Maine decides right now to control the
promise of development, Maine's greatest asset will have been
squandered, irresponsibly, and definitely."[17] Such warnings
have been given repeatedly over the last decade by professional
planners, newspaper writers, conservationists, and others concerned
with the rapid disappearance of Maine's precious coastal resources
into private control. The most recent of these was a series
of articles by Robert C. Cummings in the Portland Sunday Telegram,[18]
which outlined the results of a survey of real estate agents,
developers, town and city officials, and county courthouse records:

> Maine has probably lost its chance for significant pub-
> lic control over its 3,000 miles of coastline. Indeed,
> before the end of this decade, it appears certain that
> people will have to begin lining up before dawn on most
> good summer weekends if they want a spot at a public
> beach.
>
> This conclusion seems inescapable. Some waterfront state
> parks are already turning away visitors by noon or earlier,
> overall park usage is increasing at the rate of 20 per-
> cent a year and State Parks and Recreation Director
> Lawrence Stuart says flatly that desirable coastal prop-
> erty has practically disappeared.
>
> Campers frequently have to wait in line all night for a
> campsite to become available at Acadia National Park.
> Persons who just want to go to the beach for an afternoon
> will soon face "sorry we are filled up" problems.
>
> Dalton Kirk, supervisor of the park district that ranges
> from Eagle Island off Harpswell to Pemaquid, notes that
> admissions to Reid State Park at Georgetown are up 20 per-
> cent, despite the opening of a new park across the Kennebec
> River at Popham Beach.
>
> Kirk says that already in his region the state parks pro-
> vide the only opportunity for most people to get to the
> beach. But Reid State Park twice this season has been
> forced to turn away beachgoers when the nearly 900 parking
> spaces were filled to capacity.

And at Popham, cars are turned away almost every good
Sunday afternoon by 1 o'clock....

The state has purchased another 25 acres of mostly beach
front this summer at Popham, and Kirk believes the facilities there can be doubled eventually. But this adds only
25 percent to the region's park capacity and the number
of visitors is growing at twice this rate. Kirk sees no
possibilities of further expanding Reid State Park without destroying the naturalness of the area.

"We need to get any beach frontage that is left in Maine,"
Kirk says. But if and when the State decides to buy, it
may find little property for sale.

While pessimistic about the future status of the coast for
public use, the series stresses the importance of recognizing
the critical nature of the problem in order to avoid the same
mistakes with inland lake and mountain areas, already under heavy
pressures of speculation and development.

While Maine debates the pros and cons of oil refineries,
sulfur reduction plants and aluminum processing, a quiet
revolution in land ownership continues which promises
to bar all but the most affluent from our 3,000 miles
of ocean frontage.

...development has already progressed to the point where,
regardless of what the state does, there is unlikely to
be enough suitable ocean frontage to serve Maine and its
ever-increasing hordes of summer visitors.

Our survey reveals that Maine's coast has been sold, and
that the buyers are largely from out of state. Big
blocks remain in the hands of speculators and developers,
and while plans are being made, Maine citizens are wander-
ing at will as before, fishing the rocks, harvesting the
crops of wild berries and enjoying secret picnic spots.

But the pattern has been set. Wildland that in some cases
was sold for unpaid taxes as recently as a decade and a
half ago is about to become sites for luxury vacation and
retirement homes with shore frontage selling for up to
$100 a foot--or $20,000 for a 200 foot lot.

Much of the coastal zone is in out-of-state ownership, which
averages 45 percent in the area but reaches 75 percent in many
communities. Many real estate brokers reported that 80 percent
or more of their business had been with out-of-staters. This
boom is related to all the factors previously mentioned: increas-

ing populations, growing prosperity, and better transportation
such as the Maine turnpike and highway system that makes half
the state's coastline no more than a three-hour trip from Boston.
These factors, combined with the desire to get away from the
metropolitan atmosphere of city strife and pollution, have led
to the unprecedented demands currently placed on Maine's coastal
real estate. As a consequence, "Maine residents, the greatest
number of whom find the stakes too rich for their income, have
found themselves shut off from the sea and the wilderness by
out-of-state buyers who put up a sign before they put up a
house."[19]

Maine is not alone in facing the difficulties here described.
All of the coastal New England states are now facing serious
problems with the saturation of existing shoreline facilities.
A general inventory of coastal recreational resources for the
other New England states of Massachusetts, New Hampshire, Connecticut, and Rhode Island is included in the discussion in Section
VIII of this chapter. This inventory also reflects the critical
status of shoreline resources which points to an immediate,
urgent need to protect all the shoreline resources still available,
and to look for ways to reverse the trends of decreasing supply.
"The welfare of American society now demands that man-made laws
be extended to regulate the impact of man on the biophysical
environment so that the natural estuarine zone can be preserved,
developed, and used for the continuing benefit of the citizens
of the United States."[20]

We might ask why this has not been done in the past. The
answer lies partly in the attitudes taken toward the coastal
zone within our institutional environment. Until recently,
most states and communities were not cognizant of the coastal
zone as an environment separate from other regions of the state
and in need of special attention. In addition, there has been
a lack of cooperation and coordination among local, state, and
federal agencies, and private industries, especially where conflicts
of interest (economic or political) existed. Hence, most planning

for the use of the coastal zone has been done by bits and pieces,
in small increments, and by reacting to crises when they materialized (and usually too late for constructive action to be taken).
Prior to World War II, what planning that was done on a national
scale had objectives that "were largely resource-protection-
oriented, and the facility development which took place during
the 1930's was directed far more at providing employment than
meeting, in a planned fashion, identified outdoor recreation
needs."[21]   Such thinking was in evidence when the national park
and forest systems were established in western areas of light
population, far removed from the recreational needs of urban
centers. It seems ironic that planners would recognize the
need to preserve vast expanses of untouched wilderness in the
remote corners of the nation while ignoring the necessity of
protecting the relatively-limited coastal resources in the heart
of the country's most rapidly-expanding regions. Not until more
recent times have investigations by the National Park Service,
the U.S. Forest Service, and the Outdoor Recreation Resources
Review Commission (ORRRC) brought to light the need for a broader
concern for all issues related to satisfying the needs and
demands for all forms of outdoor recreation by present and future
generations. These studies for the first time demonstrated the
basic causal factors in outdoor recreation demand. In effect,
they found that "adequate planning for outdoor recreation required larger concerns than the biophysical environment--that
the *economic environment*--expressing the preference of society
for goods and services--and the *institutional environment*--de-
cisions about the focus and characteristics of agencies charged
with the protection of resources and the provision of outdoor
recreation facilities--were equally important."[22]

It is in this context that we identify the area of shoreline recreation to be in critical need of effective planning
and active land-use management. We will examine, within the
framework of Chapter 1, the sociology behind society's need for
outdoor recreation, the economics of shoreline supply and demand,
and the institutional aspect of coastal zone management, all in

recognition of the limited tolerance of this finite and valuable
resource to the rude invasion of man, and all in the hope that
society will perceive the problems clearly and proceed to do
something about them.

4. Summary and Overview

The purpose of this section has been to provide a general
picture of the national supply of recreational shoreline. While
a detailed inventory was not included, it is possible to draw
some general conclusions by looking at the overall situation.

The first statement we can make is that the shoreline of
New England in particular and the United States in general is
predominantly in private hands. Shore property is highly desirable
for recreational use and as long as it is available there will
be people to buy it, regardless of the cost. In every state
the patterns of private ownership and development are similar:[23]
97.2 percent in Massachusetts with high development; 94.4 percent
in Connecticut with high development; 90.4 percent in Rhode
Island with high development; 88 percent in New Hampshire with
very high development; and 98.7 percent in Maine with initially
low but more recently a mushrooming development rate. Only
in the northernmost parts of Maine are there relatively large
blocks of shoreline that remain undeveloped, and even these
are presently in the hands of speculators and developers. To
make matters worse, it is almost universally the case that competing uses preclude use of the shoreline for public recreation.
"Recreation and commerce, recreation and housing, recreation
and industry, recreation and transportation...in most cases
cannot be carried on in the same place. The practical and aes-
thetic requirements of clean water, adequate land area, safety
and pleasant surroundings, and necessary recreation developments
can rarely be assured in conjunction with commerce, industry,
housing and transportation."[24]

For years, many shore owners have permitted public access
and use of the beach and bluff areas in their possession. However, as the numbers seeking recreational pursuits in these areas

increase each year, many states are finding that their private
owners are now limiting such activity to maintain their own pri-
vacy. Hence, as the demands increase, this one part of the
accessible supply is actually decreasing. The situation is typi-
fied in the words of Pat Sherlock of the Associated Press in an
article entitled "The Best of Maine Lost to the Rest of Maine":[25]

> The mountains are still there, the Atlantic Ocean still
> crashes its surf onto the rocks as it has done since
> the Ice Age and there is still some wilderness. It's
> just a little farther away now--on the other side of
> the fence.

A second major point to be noted is the present saturation
of most publicly-owned facilities. On the Connecticut shore,
where the recreation facilities are under strong demand pressures from the dense New York-Connecticut metropolitan area,
local communities find it necessary to institute user fees,
parking charges, and other discriminatory devices to preserve
for the local residents what small amounts of shore are left
open to the public. The situation is much the same near other
population centers in New England. Beaches on Narragansett Bay,
Cape Cod, and in the Boston Metropolitan region are jammed almost
every weekend in the summer, while the beaches farther north
become more crowded each year as New Englanders search for new,
less crowded, accessible recreational areas. This trend is evidenced by the marked increase in traffic patterns this past summer leading from Boston to the southern parts of New Hampshire
and Maine.

The third and final major issue in shoreline supply is the
influence of pollution and erosion, often caused by heedless
development in ecologically-delicate areas. Pollution, usually
most severe where people are concentrated in large numbers, has
closed or destroyed beaches and presents a continuous threat
in places like Connecticut and New Hampshire, where available
beaches are scarce to begin with.

So this is the overall picture of shoreline supply: most
of the land is privately owned and developed and is becoming

more restricted to public access as the demands grow larger;
and what is left in public lands for recreation is either saturated by hordes of users or unavailable for use due to pollution
or erosion, especially near large cities. All this is to say
nothing of the future. While the demands grow at a breakneck
pace, the supply, limited to begin with, is shrinking steadily.
How can we expect to satisfy the demands of the future when we
are having trouble supplying that which is needed today? And
all this with effectively no shoreline left to do anything with!

In the next sections we develop the rationale for national
and regional concern for the problems of the shoreline through
a discussion of the needs and demands of American society for
outdoor recreation. Thus, the groundwork will be firmly established for a substantive analysis of the problem, and what to
do about it, in the remaining sections.

III. THE NEED FOR OUTDOOR RECREATION

1. Historical Attitudes

Since the earliest days of planning for outdoor recreation,
great emphasis has been laid on its value in helping cure the
ills of society. Many advocates of outdoor recreation described
parks, playgrounds, beaches, and other opportunities for recreational
activity as "veritable cure-alls which would isolate young people
from and immunize them against the delinquency, alcoholism,
prostitution, and crime that abounded in the slums."[26] In later
years, the emphasis shifted to the value of outdoor recreation
in counteracting the harmful effects of the stress and tensions
of life in an urban-industrial society. Recreation generally
came to be viewed as a major solution to the problems of mental
illness that were attributed to such tensions:

> ...people who advocated outdoor recreation were so con-
> vinced of the health-giving virtues of rural life and
> the desirability of defending rural and small-town
> America against the surge of immigrants that there was
> no need for evidence. The skeptic needed only to look at
> the slums of New York, Boston, or Philadelphia, in which

trees, grass, and fresh air were rare indeed, while crime
and mental illness flourished.[27]

Herbert Gans, the noted sociologist, has taken issue[28]
with this orientation towards a causal link between recreation
and mental health:

> ...[These attitudes were] developed by a culturally
> narrow reform group which was reacting to a deplorable
> physical and social environment and rejected the coming
> of the urban-industrial society. As a result, it
> glorified the simple rural life and hoped to use outdoor
> recreation as a means of maintaining at least some
> vestige of a traditional society and culture. Given
> these conditions and motivations, no one saw fit to
> investigate the relationship between outdoor recrea-
> tion and mental health empirically.

How then can we go about determining what relationship,
if any, exists between recreation and mental health or, in
broader terms, the general health and well-being of man in modern
times? Hopefully, the answer to this question will shed light
on some very important issues in planning for the outdoor recreational needs of American society.

2. The Individual in Modern Society

Most psychologists and sociologists would concur that the
human predicament can best be described as the task of maintaining a balance, both internally and externally, between man's
existence as an *organism* and as a *personality*. This predicament is described by Lawrence K. Frank:[29]

> So long as man lives, he must function as an organism
> through his continual intercourse with the natural
> environment, breathing, eating, eliminating, sleeping,
> and sexual functioning as a mammalian organism. Thus,
> as an organism, man is continually exposed to a variety
> of biological and psychological signals to which he is
> more or less susceptible; but, as a personality, he
> must strive to live in his symbolic cultural world,
> exhibiting the orderly patterned conduct and required
> performance in response to the symbols and rituals of
> his social order. He finds himself often "tempted"
> by these potent biological signals but continually re-
> minded by the symbols and especially by the expectations
> of other persons, of the group-sanctioned code of con-

duct he is expected to observe. This conflict is life-
long and apparently inescapable unless the individual
withdraws completely from social life in some form of
mental disorders. *A crucial problem for mental health
is how an individual can resolve this conflict without
incurring high costs psychologically and persistent
damage to his personality, and what sources he can rely
upon for strength and renewal in facing his life tasks.*
(Emphasis added)

Margaret Mead, the noted anthropologist, has posed the
same problem in more sweeping terms:[30]

There is good reason to believe that man's evolutionary
progress depends upon this ability to dream and to maintain within himself, and through his culture, a balance
between internally oriented, proprioception and externally oriented, exteroception. The disturbance of this
balance may be one of the factors which accounts for the
onset of boredom and apathy, the loss of evolutionary
vigor, and the decline of particular civilizations for
whose fall no adequate external explanation has been
found.

The significance of these statements is consolidated in the
words of Herbert Gans:[31]

Mental health is the ability of an individual as an
occupier of social roles and as a personality to move
toward the achievement of his vision of the good life
and the good society...mental health is a social
rather than an individual concept, because if society
frustrates the movement toward the good life, the mental
health of those involved may be affected.

There are considerable present-day indications that society
does tend in many ways to frustrate an individual's movement
toward the good life, and that it is increasingly difficult to
maintain the balance necessary for well-being as described
above. The characteristics and intensity of the emotional
stresses and strains of modern life have been stated (and sometimes overstated) by many writers.

Many of the facts of urban life are inescapable. The air
environment is often polluted by smog, gaseous effluents, particulate matter, and other contaminants; highways are jammed
with traffic; and noise and crowds are everywhere. The sociological effects on man of such an environment have been discussed

by Lawrence Frank:[32]

> We are beginning to realize that this urban crowding and
> enforced contacts with strangers, plus the continual sen-
> sory overloads, may have serious impacts on human person-
> ality. Man is well prepared to deal with sudden emergen-
> cies, to cope with physical threats and actual situations
> that release his energies for overt activities, but he is
> less well equipped to bear prolonged strain, to be unre-
> mittingly alert and vigilant, under sensory overloads.

There is no doubt that the pollution, congestion, noise, social
ills, and just the hectic pace of the urban environment detract
from the well-being of those who live and work in the metropolitan areas. These "sensory overloads" have particularly
severe effects on the low-income, less mobile groups that now
dominate the central cities. Here the sensory overload is compounded by extreme crowding and oppressive living conditions,
by widespread nutritional inadequacies, and by the frustrations
of unemployment, drug addiction, and high crime rates.

Having established that health can best be understood as a
product of the interaction between an individual and the total
physical and social environment that he experiences, and, recognizing some of the impediments to the maintenance of a healthy
sociological balance in this interaction with present-day society, we must now ask: what part can outdoor recreation play
in helping the individual maintain this balance so vital to his
mental health and physical well-being?

3. The Role of Outdoor Recreation

Recreation has always been a prime objective of life, even
since the times of the early Greeks of the fifth century B.C.
Today, most Americans, when given an opportunity to diminish
their sensory overloads through a change of routine, "will spend
a summer afternoon in a suburban backyard around a barbecue, in
a city park, or at the nearest swimming pool or beach. Given
the chance and the means for a weekend or a vacation away from
home, they will take to the country, the mountains, or the seashore."[33] It seems undeniable that the opportunity to secure
and the ability to participate in satisfying leisure behavior

are fundamental ingredients in any determination of the "good
life." Satisfying leisure behavior, according to Gans, is "the
emotional relaxation, reduction of fatigue, restoration of
energy lost elsewhere, and general recreation without ill
effects."[34]

The question now is: what is the role of outdoor recreation in relation to satisfying leisure behavior? No one can
deny that serious emotional and nervous tension exists today and
that many people find release in outdoor recreational activity.
"But it is by no means clear that everyone, or even a majority
of persons, suffers from severe strains and stresses; moreover,
a substantial proportion of the population apparently rarely or
never engages in outdoor recreation....Although much is made of
the increase in tension and strain, yet it is a fact that no
comprehensive continuous effort has ever been made to measure
these factors..."[35] So, while there are obvious positive benefits to be derived from outdoor recreational activity by many
persons, it should not be pointed to as a panacea for the many
ills of society. Herbert Gans has presented the most incisive
approach to the issue.[36]

> I am saying that *leisure and recreation are a constitu-
> ent part of mental health*, but they cannot by themselves
> bring about mental health, cure mental illness, or pre-
> vent it...they are essential and desirable, but they
> are not so important as economic opportunity and secu-
> rity, positive family life, education, the availability
> of a variety of primary and secondary group support, and
> the like...the recognition of the limited significance of
> outdoor recreation in the treatment of personality dis-
> orders should not blind us to the potential significance
> of it for developing and sustaining healthy personalities.
> Indeed, we may find that recreation, especially outdoor
> recreation, *provides one of the most promising approaches
> to the elusive goal of mental health as a form of "primary
> prevention" of mental ill health*. In and through out-
> door recreation the individual, especially in early life,
> may develop the self-confidence, the elasticity, and
> spontaneity for action and expression of feelings which
> will enable him to cope with city living and indoor
> working, while maintaining his physical and mental health.
> (Emphasis added)

Hence, we should view outdoor recreation for what it really
is: not a solution or counteraction of the evils of urban-industrial society, but an enjoyable form of leisure behavior that
appears to contribute to mental health in that it offers "a
change from one's daily patterns and an opportunity to find
self-identification and personal achievement in ways that the
daily patterns do not afford."[37]

4. Conclusion--The Approach to Planning

We have concluded that the arguments for the psychological
and emotional *need* for outdoor recreation may have been overstated. Each individual takes a different view of recreation,
depending on his preference and personality, is conditioned by
his physical and economic development, and is influenced by his
age and sex. From this we can see that the collection of more
extensive data on leisure behavior is immensely important. "If
we can discover what needs and aspirations people are trying to
fulfill and can recognize what may be blocking or frustrating
their quest, we can understand better what provisions to make
for future recreation. Also, we may find some clues to the
meaning of outdoor recreation for the individual personality
and its significance for mental health."[38]

How then are we to plan for outdoor recreation? It is
clear now that this presents a wide variety of sociological
questions of long-term policy and many subtle problems not easy
to define or resolve. Yet it seems undeniable that recreational
activity has great social significance and personal value for
millions of American citizens. While the empirical evidence
is relatively sparse in support of the case for the psychological
and emotional need for outdoor recreation, it is clear that
the *demand* for this type of activity is very strong and is rapidly
increasing:

> ...to ask whether outdoor recreation is important to
> the mental health of Americans is, in one sense, tanta-
> mount to asking whether the full and rich life is impor-
> tant; and the answer of course is clear...the degree
> of crowding at our parks, our ski slopes, beaches, pic-

nic sites, and even our mountain trails is clear evidence
of the popular response to this question.[39]

This suggests that the best way to plan for recreation is
to adopt a *user-oriented* approach that will provide the recreational facilities that are presently used and preferred by those
seeking satisfying leisure behavior. Having recognized this, we
now turn to a look at the patterns of recreational demand in
this country, with a focus on the New England region. There is
unanimous consent that on the basis of these trends, demands
for outdoor recreation in the future will far surpass those which
we have experienced to date. Also, it has become clear that--as
indicated by most studies that question people about their lei-
sure-activity preferences--the biggest demands will be for water-
related activities, especially swimming. Hence, in the next section, we apply this approach in determining the user demands
for outdoor recreation, with special emphasis on the shoreline.

IV. THE DEMANDS FOR OUTDOOR RECREATION

1. Basic Trends

At this point it is clear that outdoor recreational activity can be considered an important component of a full and
well-adjusted life for most Americans. Thus, it should come
as no surprise that the demand for such activity is surging,
spurred by increases in the causal factors of population, disposable income, leisure, mobility, education, and overall standard of living. The Outdoor Recreation Resources Review Commission, in a report[40] to Congress in 1962 entitled "Outdoor
Recreation for America," noted and documented these causal
factors and their influence on recreational demands. It was
the conclusion of this report that as the levels of these factors rose, the growth of outdoor recreation demand would accele-
rate even faster, and in a sustained fashion, than the net increase in population:

> Whatever the measuring rod...it is clear that Americans
> are seeking the outdoors as never before. And this is
> only a foretaste of what is to come. Not only will there

be many more people, they will want to do more and they
will have more money and time to do it with. By 2000
the population should double; the demand for recreation
should triple.

Having noted the increasing trends in the principal socioeconomic variables affecting outdoor recreation, the prospect
for future demands is clear.

> The indications are imposingly those of a more-so
> society. Attendance and use figures for outdoor areas
> are already reflecting the trends of related factors
> and are rising at continued high rates. National park
> attendance rose from about a million in 1920 to 102
> million in 1964. Total state park attendance increased
> from about 69 million to 285 million over the years 1942
> to 1962. Some areas, particularly those which are
> water-oriented, are experiencing even higher rates of
> increase in use. In view of the trends in recreation
> participation and in the factors having a direct rela-
> tionship to outdoor recreation, greater pressures on
> recreation resources seem inevitable.[41]

Dr. Marion Clawson, in an article entitled "The Crisis in Outdoor Recreation,"[42] concluded that the projections of these
principal factors to the year 2000 point to a *tenfold* increase
in the demand for outdoor recreation from 1950 levels. A report of more recent survey information on recreation trends up
until 1965 has indicated that "present and anticipated increases in major summertime outdoor recreation activities far
surpass predictions made by the ORRRC in 1960."[43] This study
predicted that by the year 2000 participation in the major forms
of summertime outdoor activities will be four times greater than
it was in 1960.

Having established some generalized trends in overall recreation demands, our next task is to look at the present and projected patterns of those demands by examining the *participation
rates* for various outdoor recreational activities.

2. The Patterns of Demand

The patterns of demand as expressed in participation rates
and user days, the most common indicators of recreational activity, are shown for the United States (1960) in Table 3.2.

	US	NE	US	NE	US	NE
	% participating		days/person		days/participant	
Picnicking	53	57	2.14	2.81	4.0	4.9
Driving for pleasure	52	54	6.68	7.23	12.7	13.4
Walking	53	43	4.34	6.46	13.1	15.1
Attending sports events	24	22	1.32	1.15	5.5	5.2
Attending concerts outdoors	9	13	.21	.33	2.4	2.5
Swimming	45	53	5.15	6.82	11.5	12.9
Playing sports	30	34	3.63	3.91	12.3	11.6
Fishing	29	21	1.99	1.76	6.8	8.5
Boating	22	21	1.22	1.38	5.5	6.7
Canoeing	2	3	.07	.09	3.0	3.1
Sailing	2	2	.05	.06	3.0	2.5
Waterskiing	6	4	.30	.29	5.1	6.5
Bicycling	9	9	1.75	1.47	19.4	16.3
Camping	8	5	.46	.33	5.7	6.9
Hunting	3	2	.19	.22	5.6	8.9
Horseback riding	6	4	.42	.29	7.5	6.8
Hiking	6	7	.26	.28	4.4	4.2
Nature walks	14	15	.75	1.14	5.2	7.5
Mountain climbing	1	2	.04	.06	3.7	3.6

Participation rates for the United States and the Northeast
(Maine, New Hampshire, Vermont, Massachusetts, Rhode Island,
Connecticut, New York, New Jersey, Pennsylvania) during June-
August 1960.

Percent of persons 12 years and over participating
Days of activity per person
Days of activity per participant

Source: "National Recreation Survey," ORRRC Study Report
         No. 19 (1962).

Table 3.2   Patterns of Demand in the United States
            and the Northeast Region

These indicators are listed for various outdoor activities and
comparisons are made between the averages of the entire United
States and the Northeast region.

The first major trend of note is that Americans most frequently participate in simple activities that are usually independent of age, income, education, or occupation. Driving and
walking for pleasure, swimming, picknicking, and sightseeing

lead the list of outdoor pursuits in days of activity per person.
Driving for pleasure is the most popular and, together with
walking for pleasure, accounted for about 33% of the total U.S.
activity days per person for the period in question, and 37% in
the Northeast. Walking for pleasure is very popular in the
urban Northeast even though cities often lack safe, scenic pedestrian areas free from annoying air and noise pollution. Walking
is an important type of recreation for older citizens and for
those with infirmities. Although nature walks are of a low preference, those in the Northeast participate at a rate twice that
of other regions. Sightseeing ranks among the highest desired
activities especially for weekends and vacations when more time
is available for longer trips than on weekdays. This has particular significance for New England where tourism is a major factor in the economic posture of most states. Picnicking is enjoyed by 57% of the people in the Northeast and is frequently
combined with driving for pleasure on day outings.

A second notable trend is the generally higher level of participation rates and user days in the Northeast in the most
popular activities such as picnicking, driving, walking, swimming
and playing sports. In metropolitan areas of this region, more
people spend more time in these activities even though in the
inner cities (where there is most need for more outdoor recreation) one finds the lowest rates of participation associated
with low-income and poorly-educated people living in oppressive
surroundings. Outdoor recreation does not play an important
role in the leisure time of these groups due to the lack of
nearby facilities and the lack of money and adequate transportation to get to more distant areas. Both of these observations
indicate that outdoor opportunities are most urgently needed
near metropolitan areas; yet this is where available land is the
scarcest. It is probable that at least 80 percent of the total
population will live in these urbanized areas by the turn of the
century. These people will have the greatest need for outdoor
recreation, yet their need will be the most difficult to satisfy
since urban areas have the fewest per capita facilities and the

greatest competition for land.

A third major trend is the pervasive attraction of *water-oriented* activities.[44]

> Most people seeking outdoor recreation want water to
> sit by, to swim and fish in, to ski across, to dive
> under and to run their boats over. Swimming is now
> one of the most popular outdoor activities and is likely
> to be the most popular of all by the turn of the cen-
> tury. Boating and fishing are among the top 10 acti-
> vities. Camping, picnicking, and hiking, also high on
> the list, are more attractive near water sites.

Swimming, fishing, boating, canoeing, sailing and waterskiing
accounted for 26 percent of the total U.S. user days per person
reported in Table 3.2. In this regard, New Englanders lead the
nation in per capita participation in water-related outdoor recreation, as shown in Table 3.3. Swimming seems to have special
importance to urban dwellers since 49 percent of the metropolitan population (versus 38 percent of nonurban dwellers) participated in the activity. In the Northeast, 53 percent of the
population swims. For the U.S. as a whole, 17 percent of those
not participating expressed a preference for swimming. This
points to an extensive need for swimming facilities to be

Activity	New England days/yr	U.S. average days/yr
All swimming	11.53	6.84
Ocean swimming	3.11	1.58
Fishing	3.05	2.26
Motorboating	2.71	1.56
Waterskiing	.75	.42
Sailing	.62	.16

Source: *1965 Survey of Outdoor Recreation*, Bureau of Outdoor
Recreation, U.S. Department of the Interior

Table 3.3 Participation in Water-Related Activities

provided close to demand centers, especially in urban areas,
where coastal beaches are generally already used to capacity.
In terms of attendance, the beaches of New York (Long Island),

Maryland, Virginia, Massachusetts, Florida, and California, all
centers of large urban populations, are the most heavily used
in the United States. The 1965 survey (Reference 43) reported
that swimming, ranked second at that time in user participation,
was becoming so popular that it will be our number one outdoor
recreation activity by the year 1980! In addition, boating and
other water-related activities will continue to increase as
long as access points and suitable water areas are in adequate
supply. In some areas there are so many boats at anchor that
room for turn-arounds is fast disappearing. In Rhode Island
alone, three hundred new pleasure boats are bought annually,
all in need of docking accommodations.

A final trend of importance to be noted here is the great
demand for activity close to home. People seeking outdoor recreation do so within definite time patterns that can be classified as day outings, weekend or overnight trips, and vacations.
The most frequent of these is the day outing, which can presently be considered as the fundamental time unit of outdoor recreation. Most indications are that, at the present time, people
will drive one way about two hours, a distance that varies from
30 miles to as much as 90 miles, for outstanding recreation
sites like ocean beaches or scenic campgrounds. For the weekend
or overnight outing, the median travel distance is about 90 to
125 miles. While many vacationers will travel many miles on
week- or two-week-long vacations, by far the greatest demands
are placed on the facilities serving daily and weekend outings.
Hence, pressures are greatest within about 125 miles of metropolitan centers, with maximum demands at those facilities in
close proximity to the central cities. For example, in 1954,
the Massachusetts Department of Natural Resources reported:[45]

> ...80 percent of the ocean beach capacity lies within
> the Metropolitan Parks District, where 2 million people,
> more than 40 percent of the State's population live.
> Within this district, where the beaches can accommodate
> 15 percent of the resident population, use on peak days
> taxes their capacity heavily.

The situation is much worse now in 1971. All this points to

the great importance of "providing outdoor recreation facilities close to where people live so that individuals of all ages
can go frequently, as contrasted with the occasional longer
trips and annual vacation pursuits."[46] Hence, today's problems
"do not center on the acquisition of the unique and dramatic
resources for the public, but on the broad availability of outdoor recreation for everyone and often; nearby open areas for
weekend visits by moderate-income urbanites are more character-
istic of our recreation needs than the annual trip to a faraway
area of unforgettable beauty by the fortunate persons who can
get there."[47]

    3. <u>Factors Affecting the Demand for Shoreline Resources</u>

    The enormous recreational demands for shoreline resources
are conditioned and directed by three important factors, as
described by the ORRRC[48] in 1962: 1) the *type* of shoreline,
2) the *availability* of the area for recreational activities,
and 3) the *accessibility* of the area to those desiring shoreline
recreation.

    There are three *types* of shoreline: *beach*, *bluff*, and *wet-
land*. Of the three, beaches are by far the greatest in demand
because of the wide variety of recreational activities that
they support. Bluff shores, characterized by bank, bluff, or
cliffs immediately landward of a narrow beach, provide "a marine
environment, scenic values of a high order, and frequently the
isolation many outdoor recreation seekers prize so highly."[49]
As such, they are in demand chiefly by hikers, campers, and
sightseers who form a sizable group but are small compared to
the hordes who flock to beaches. Wetlands are characterized
by tidal or nontidal marsh. These shore areas are least in demand
as recreational areas, although they are very attractive to
developers who would fill in the marshes for commercial building.
Yet of all the shore areas, the wetlands are probably the most
valuable in the ecological sense because of the wide variety
of fish, plant, and wildlife that they support.

    The second factor in the demand for shoreline resources is

availability, or the absence of restrictions that inhibit the
use of a particular area by would-be recreationalists. This
restriction could be due to private ownership, high fees, lack
of adequate support facilities such as parking, or pollution.
In general, "the only beaches widely available to the public
are public beaches, and even some of these are restricted.
For example, some municipal beaches admit only bona fide citizens of the municipality. Others practice some form of segre-
gation"[50] such as exorbitant parking fees for nonresidents.
Especially distressing is the fact that of all the coastal
resources as of 1960, only about six percent are public recreational areas, while the other 94 percent are not available for
public recreation due to private and military ownership.

The availability of recreational activity also has a definite sensitivity to the quality of the environment in which
that activity takes place. "The quality of water is as important as the amount of surface acres, miles of banks, or location. Polluted water in the ocean, a lake, a river, or a reser-
voir is of little use for recreation. Pollution by human or
industrial waste is only one aspect of quality which conditions
the available supply. The silt load, the bottom condition, temperature, and aquatic plants also affect the usability of water
for recreation."[51] Yet, in most major cities, pollution has
destroyed the availability of otherwise ideal recreational opportunities, just where they are needed most (Boston Harbor,
Lake Erie, etc.).

Finally, the demands for coastal activity are conditioned
by the *accessibility* of available and suitable shoreline resources. Accessibility of a recreational area to any given user
depends in part on that user's income and mobility. While the
upper-income urban groups can often afford either second homes
in some distant recreational areas or extended stays at re-
sorts, the great majority of people in the middle-income brack-
ets prefer to vacation within a maximum of approximately 90
miles of the urban areas, while low-income residents of the

central city often are not able to leave the confines of the
metropolitan area at all. Hence, the enjoyable use of coastal
recreational resources for these groups is closely linked to
the availability and suitability of beaches that are within
(or very near) the metropolitan area itself. Yet, it is in
these areas that the demands from the competing uses of private
housing, commercial and industrial development, and transportation
are all extremely heavy, while the problems of pollution are
particularly severe.

4. Summary

We have seen in this section how the demands for outdoor
recreation are great--especially for water-oriented activities--
and will inevitably increase rapidly with the upward trends in
population, leisure, income, and mobility. The combined multiplying effect of these trends--more per capita leisure, mobility, and income applied to a population expected to double between 1960 and 2000--is projected to be a *tripling* in the demand
for outdoor recreation from 1960 to 2000, while much of this demand will be concentrated in the densely-populated metropolitan areas. We have also noted that shoreline resources have a
particular attraction for large numbers of people, while their
demands are conditioned by the type of shoreline, its accessibility, and its availability.

From this outline of the proportions of future demands
for outdoor recreation, we can draw some clear implications
as to the future of *shoreline* recreation. With continuing increases
in population, leisure, income, and mobility, the demands for
shoreline recreation should *triple* before the turn of the century.
Such an increase is staggering when we consider that *our public
coastal facilities are already filled to capacity*, while there
is no room left for expansion through acquisition and development
since the remainder of the shoreline is already owned for private
development! Each summer we feel the pinch of this disproportion-
ate situation of shoreline supply and demand as hordes of recrea-
tionists crowd the beaches, especially near the cities, along the

entire perimeter of the nation.

All this points to the great value that Americans place
on outdoor recreation, especially that which is water-related.
In the next section, we will find that further exploration of
this value will give us a firm rationale to serve as the basis
for the analysis in section VII.

V.   THE VALUE OF SHORELINE RESOURCES TO AMERICAN SOCIETY

The fact that the demands of American society for shoreline
and other outdoor recreational activities are so great is clear
indication that we attach significant value to this aspect of
our experience. This value is manifest in a number of forms,
the most important of which are 1) the intrinsic value to the
health and well-being of all citizens, and 2) the concrete
economic value to regional communities. We shall explore both
of these.

1.   The Intrinsic Value of Coastal Resources

The preceding discussions on the great social significance
of outdoor recreation and the fantastic demands that we now see
for shoreline activities point to the unique and intrinsic value
of our coastal zone as a recreational resource. This value has
been pointed out by the ORRRC: "Of the many outdoor recreation
'environments,' mountains, seacoasts, deserts, and woodlands,
the shoreline appears to have an unusually strong appeal for
Americans."[52] This is true because of the wide variety of easy,
active forms of recreational activity that the shoreline affords.
This wide variety includes swimming, skindiving, beachcombing,
motorboating, sailing, canoeing, waterskiing, and fishing. Many
other activities, such as picnicking, camping, sunbathing, and
walking are greatly enhanced by proximity to the ocean. Beach
shoreline, in most cases, offers the cheapest and most enjoyable
recreation uses for large numbers of people.

> Going into the surf is fun whether one swims or not.
> It is not necessary to be a mountain climber to take
> walks along the beach, and beachcombing is an activity

that appeals to everyone from toddler to octogenarian...
here, land and water are easily accessible; the violence
of breaking surf and the warm safety of relaxing sands
are but a step apart; the stimulation of the foreign
environment of the water and the relaxation of sunbathing are nowhere else so easy of choice. Physical
sport and mental relaxation are equally available.[53]

An additional use of coastal areas, and probably the most
widespread, is for esthetic enjoyment, especially along bluff
shoreline.

Tourists from the interior states are always eager to
view such sights as ships coming under the Golden Gate
Bridge into San Francisco Bay, the lovely solitude of
Fort Sumter as it rests seemingly impregnable in
Charleston Harbor, and the parade of ships in and out
of New York Harbor. Attractive scenic vistas are not
for the tourists alone, but hold a certain magnetism
for residents of the coastal cities as well. One has
only to scan the real estate advertisements to realize
the premium value on waterfront or waterview lots.[54]

All these values of the shoreline are magnified by its
accessibility to large populations. "This unique combination
of available resources in close proximity to large population
centers offers an unparalleled recreational opportunity for many
people who could not afford to travel far from their homes,"[55]
and as such is an invaluable asset of this nation.

The coastline has great value in another important sense.
Although man is a social being, performing social activities
such as recreation, he is also a biological organism, "one
species among many who depend upon each other and upon the
natural environment for their organic needs...his very survival
depends upon the intricately complex, ecological balance among
all plants and animals within their respective geologic and
climatic environments."[56] This points to the unique value of
the coastal zone as an ecological system and as a basic element
in the environmental life cycle of all living things. The many
forms of fish and wildlife found solely in the coastal and
estuarine zones are an integral part of this ecosystem, together
with all other life-forms that exist in the beach, bluff, and
wetland areas of the shoreline. There is a clear and pressing

need to preserve the vitality of all such ecological systems,
at the very least until man can determine their ultimate importance as a component part of his own life cycle and those of
other forms of life on this planet. "An awareness of man's
place within the total natural environment is clearly essential
to the understanding of the very nature of man and to his best
adaption to this environment."[57]

2. Impact of Recreation Spending on the New England Economy

Recreation and tourist spending are mainstays in the economies of the New England states. In Maine, New Hampshire, Vermont
and Massachusetts, this "industry" stands as the second largest
source of revenue. Recreation in New England is a booming business and is expected to grow rapidly with increases in population, leisure time, and income, aided by the higher mobility
brought on by better roads and other transportation facilities.

Nationally, expenditures on recreation and travel have
been increasing at a substantial rate. In 1964, 23.8 billion dollars were spent in the U.S. on recreation, a 500 percent increase over expenditures in 1940, while the total population
increased by 45 percent during the same period.[58] In 1960,
leisure-time spending was 12 percent of all personal consumption expenditures. Although outdoor recreation is only one of
many kinds of leisure behavior, it accounted for one-half of
this spending, or 6 percent of all expenditures. In addition,
one-half of all outdoor recreation spending occurred while away
from home communities.[59]

Table 3.4 shows the total recreational trade revenues generated in each New England state in 1963. In Maine, only revenue
from forest products exceeds that generated through recreation
and tourism, which contributes 400 million dollars per year.[60]
From 1958 to 1962, a short span of four years, there was a 58
percent increase in hotel and motel receipts. Per capita receipts of 56 dollars are well above the U.S. average of 48
dollars.

State	Amusement & Recreation Services (no movies) (thousands dollars)			Hotels, Motels, Tourist Courts, Camps (thousands dollars)			Per capita receipts Total trade* (dollars)
	Total estab.	Total trade	% change from 1958	Total estab.	Total trade	% change	
Me.	540	12,965	55	1402	41,730	14	56
N.H.	401	20,451	89	1146	38,747	19	91
Vt.	219	11,994	119	749	26,340	50	96
Mass.	2324	100,581	34	1376	127,103	30	42
R.I.	439	22,076	34	172	11,715	33	38
Conn.	1009	35,090	48	562	44,000	27	29

*U.S. average per capita expenditures: $48.

Source: U.S. Dept. of Commerce, Bureau of the Census, Census of Business, Vol. VI, 1963.

Table 3.4   Service Trade Revenues in New England - 1963

In New Hampshire, vacation spending was responsible for 249 million dollars, or 20 percent of total income during 1960.[61] Although total spending by vacationers was 146 million dollars, the initial recipients spent money on wages, rent, supplies, etc., so that final expenditures, through a multiplier effect, were calculated to be 249 million dollars. Receipts from amusement and recreation increased 89 percent,[62] while per capita receipts were 91 dollars. Spending by out-of-state vacationers, who owned 60 percent of the seasonal houses in 1960, creates 25,000 additional jobs during the summer months and provides the principal source of income for many seaboard towns.

In Massachusetts, income from tourism in 1968 was over one billion dollars, the state's second largest source of revenue. In eastern portions of the state, recreational spending has increased only 4.5 percent a year while U.S. spending on recreation increased 20 percent and travel by 14 percent.[63] Employment in recreation and tourist industries is becoming less important in relation to total nonmanufacturing employment and is also falling absolutely. This is happening because the supply of facilities has not kept pace with the demand and facilities have become

rundown and overcrowded. Historic sites are the most popular
attractions but they are poorly promoted, poorly coordinated,
financially weak, and ill-equipped for a greater influx of tourists.
Hence, many residents find activities in neighboring states
much more desirable. Even so, tourist flows of 12 million per
year are expected to increase to 24 million by 1990, based on
existing traffic patterns and Massachusetts' share of the national
market. Cape Cod is economically dependent on the resort business
and expenditures there are to increase from 92 million dollars
in 1960 to 227 million by 1980. The main problem in Massachusetts
is clearly saturation of existing facilities.

In Rhode Island, the shore industry has been growing by
2 million dollars per year since 1952 when revenues totaled
18 million dollars from tourist spending.[65] By 1970, 45 million
dollars is expected annually in such revenues. From 1958 to
1962, this state experienced a 34 percent increase in receipts
from amusement, recreation services, and lodging facilities.
Per capita receipts totaled 29 dollars in 1963.

VI. ANALYSIS OF THE ALLOCATION OF SHORELINE RESOURCES

The allocation of coastal resources in this country has
always been determined within the *economic* environment of the
private marketplace and the *institutional* environment of *local
political decision-making*. In the analysis of these mechanisms,
we can determine what factors have led to the present shortage
of shoreline supply for outdoor recreation uses.

1. The Economic Environment

In the economic analysis of Chapter 1 we saw that the private marketplace is the mechanism through which society exercises the choice between alternative allocations of scarce resources. If certain basic conditions are met, there will exist
a set of market prices such that profit-maximizing firms and
benefit-maximizing consumers who respond to those prices will
automatically direct the economic system into the most efficient
(consistent with the values of society) allocative position.

However, even the most loyal defenders of the competitive market
system will admit that there are circumstances in which markets
fail to provide worthwhile outputs and underproduce others. We
have investigated when and why private markets might not work
well in order to determine steps that might be taken within the
institutional environment to correct for misallocations of
scarce resources. We found that the characteristics of some
goods, which we call "public goods" point to the breakdown of the
allocative price mechanisms since they all involve violations of
the necessary conditions of a properly-functioning market. The
crucial point is that frequently the total opportunity costs to
society are *not* reflected in the price of those goods. Although
the social benefits of having an individual consume/produce (or
not consume/produce) a particular commodity may exceed his private benefits, he will base decisions only on his private benefits. *The private market, left alone, tends to produce too many
private goods and too few public goods.* This happens because the
public goods are *undervalued* within the private market and are
unable to compete on an equal footing with other goods in the
allocation of scarce resources. For this reason, government must
step in and initiate some form of collective action in order to
maintain social balance and achieve an efficient resource allocation consistent with the overall goals and values of society!

We are now in the position to make the connection between
shoreline recreational resources and their allocation in a private market economy. In the discussions of Chapter 1, we noted
that society often places a high value in charging *collective*
institutions with the responsibility of allocating the scarce
resources we have classified as public goods. Examples of such
goods are police and fire protection and public education.
The high societal value placed on such goods stems from the
idea that everyone in a democracy has certain inalienable rights
to them and that everyone, regardless of income, should derive
equal benefit from their provision. The tremendous importance
attached to such goods demands that they be allocated with extreme care to ensure that all members of present and future

generations can derive maximum benefit from their use.

In this context, it seems clear that the American
shoreline should be considered a public good in every sense of
the word. We have established that it has an intrinsic value to
society as a recreational resource in that everyone in a democracy has an inalienable right to derive equal benefit from the
value of shoreline recreation to his physical, mental, emotional,
and general well-being.

> A new slogan, declaring that recreation is the fifth
> freedom that we now urgently need to gain and enjoy the
> other four freedoms, might elicit a nationwide response
> and a reaffirmation of our traditional goals and historic
> aspirations.
>
> Seen as an indispensable, vitally imperative need in the
> great movement for *human* conservation, we can say that
> opportunity for outdoor recreation today is also an un-
> deniable human right in a democracy...no one should be
> deprived of outdoor recreation through which individuals
> can make human living more significant and fulfilling,
> more conducive to the realization of their human poten-
> tialities and attainment of our enduring goal values.[66]

Hence, the unique nature of the coastal zone as a recreational
resource, as an esthetic attraction, and as a fish and wildlife
habitat important in their ecological links to man make it an
invaluable component of the human environment. As such, we
identify the shoreline as a public good of the highest
value and in need of most careful allocation. Yet, in the past,
it has been precisely this value that has been subordinated
to the more immediate economic needs and demands of society. It
is important to discover why this has happened, why there has
been a serious misallocation by the private market of this
scarce and uniquely valuable natural resource, and how we can go
about correcting previous errors.

Historically, those uses that could pay the highest prices
for the land have preempted most of the shoreline. These uses
have most frequently been for industrial and commercial development, housing, and private recreation, all of which have for a
long time been well-established in the competitive marketplace.

The allocative mechanisms of the market have functioned well with regard to the distribution of coastal land among these competitors. Unfortunately, public recreation has never been able to participate effectively in the competitive process. Hence, the bids for land from these uses far outstrip those of public recreation and have led to the supply situations discussed in previous sections, i.e., most of the shoreline is in private hands.

Competition for coastal land is particularly strong near the metropolitan areas where the demands for private recreation, housing, commercial development, and industrial development are all heavy. This results in severe escalation of shoreline land prices, even at greater distances from urban regions where the competition is usually between public and private development for recreation. Even here the demands of private parties for recreational shoreline have forced the prices beyond the reach of many local economies. The Bureau of Outdoor Recreation has pointed out this keen competition between individual developers and public agencies for prime recreation lands, especially those that are water-oriented. In a 1967 report[67] on land-price escalation, the Bureau reported that land values were generally rising, on the average, from five to ten percent annually, while the prices of lands suitable for public recreation were rising at considerably higher rates. As an example, the report cited an initial appropriation ceiling of 14.0 million dollars established by Congress for the acquisition of land for Point Reyes National Seashore in California that was subsequently raised to 57.5 million dollars, an amount more than four times the original authorization.

As long as there was plenty of available shoreline to satisfy all the demands from competing uses while still providing adequate opportunities for those seeking recreational activity, there was no perceived need to reject the private market as an allocative system. Even today, the market is functioning in a predictable way: as the supply gets smaller in the face of heavy demands, the price goes up. However, it is

now clear that the conditions necessary for an optimal allocation of resources consistent with the values of society are no
longer fulfilled in the operations of the private market regarding the coastal zone. One difficulty, clearly involved with
some aspects of shoreline allocation, is that it is often impossible to put a price on certain values, much less find a way to
translate these values into revenue. For example, consider the
difficulty in trying to determine the value (in dollars and
cents) of bluff shoreline as an esthetic attraction. Conceivably
a developer could provide coastal roadways with scenic vistas
and charge user fees; but the uncertainty in setting a fee based
on willingness-to-pay and the prospect of little or no short-
term return on a large investment makes this highly unlikely.

     Another relevant point is that people will often misstate
their values, depending on whether or not they think something
will be provided anyway. This would come into play if a state
were to try to decide on a tax to be levied to support the provision
of recreational facilities. Most people, when asked, would
understate their values to minimize tax payments, and the resulting
tax revenues (if based on what the people said) would be too
low to effectively finance proposed programs. This fact, together
with the large numbers of people who must be polled, lead to
high transaction (contracting) costs in the gathering of such
information and seems to provide an insurmountable obstacle
(in many cases) to the gearing of any system that allocates
public goods on an individual's willingness to pay. On the
other hand, the development of the shore as vacation home sites
would provide an immediate return on investment that is determined
by a well-defined price. The same is true of most other uses
for the shoreline: hotels, motels, factories, and power plants
all begin to show relatively high return on investment shortly
after they are put into operation. Public recreation in general,
such as ocean swimming at a beach, ranks low in this regard;
unless a developer decides to provide facilities on a large
scale (such as amusement parks), there is little chance that
public recreational uses can compete with private, commercial,

and industrial development. Yet, use for public recreation
may well represent the largest value to overall society although,
regrettably, this value is the least quantifiable. As long
as a public good such as shoreline recreation is not forced
to compete with other uses, there is no need for any valuation
at all. In present-day circumstances, however, there is not
much shoreline left for development, and all uses must compete
under the same ground rules. Hence public recreation, undervalued
because of the difficulties outlined above and lacking in any
mechanisms to discover and translate true values into revenues,
has been priced out of the market.

Another way to view this problem is from the standpoint
of *side effects* that accrue to future generations. We have
seen that public recreational uses, undervalued as a public
good in the private marketplace, cannot compete effectively
with private development for commerce, industry, or housing,
while these activities almost universally deny other uses, especially public recreation. Now, this consumption and subsequent
exclusion of the shoreline by private development gives rise
to the side effect of lost opportunity to future generations
to use this resource in its unique capacity for recreational
activity. Under normal conditions (price fixed at proper level),
this exclusion is an indication of a properly functioning market.
However, in the context of undervaluation, exclusion results
in an allocation of the good in a way that is inconsistent with
the overall benefits and values of society. Since information
on the true value of the resource for recreation is hard to
determine within the framework of the price system, and since
the transaction costs of transferring information of this kind
(even if it were available) into revenues are prohibitive, the
market mechanism fails to provide reasonable competition in
which recreation uses could participate. If recreation values
could be imputed by the market, it is likely that the private
costs for the shoreline would be astronomical (even relative
to today's prices) and would greatly alter the patterns of

consumption. But since they are not, since the true costs (including the externality of lost opportunity for public recreation
in the future) are not generally included in the price of shore
land, the public must bear these social costs while the pattern
of private consumption and development continues unchecked.
These factors point again to the identification of our natural
shoreline as a public good in sore need of a means of allocation
other than that provided by an inadequate private market.

This completes the description of the *economic environment*
within which the allocation of shoreline resources takes place.
But this is only half the picture, since historically the private market has been subject to regulation within the institutional environment of *local political decision-making*. This
is the next topic for discussion.

2. The Institutional Environment

The institutional environment of the American shoreline
is made up of a large, diverse group of governmental units having
some jurisdiction and control over varying amounts of coastal
property, usually through local ownership or state authority.
These units establish the political and legal constraints under
which the private market operates. From this complex structure
of fragmented municipal, state, and federal responsibilities
for the management of coastal affairs stem the barriers to effective resolution of conflicts among different interests competing for the use of the same resource. *The planning of public recreational services has traditionally been carried out
through the process of local political decision-making while
the state of the techniques by which decisions are made is
depressingly low.* At all levels, shoreline recreation has encountered the common nemesis of all public services--"the
stifling effect of jurisdictional boundaries which, by a curious
osmosis, permits the diffusion of problems throughout the region,
while blocking any corresponding flow of governmental responsibility."[68] This points to the natural consequences of fragmented political control. While the shoreline is obviously

no respecter of political boundaries, its control is distributed
among discrete units. These units are virtually forced to act
inefficiently with respect to the values and interest of the
overall society of the region due to 1) the nature of the demands,
2) the irregular distribution of the resources, and 3) the particular economic and political context within which each unit
makes decisions.

The aspect of demand that contributes to an inefficient
allocation of coastal property by local communities has to do
with the ever-increasing mobility that brings hordes of recreationalists to any richly-endowed area within an expanding radius
of urban areas. This, combined with the irregular distribution
of these areas within the region, results in a steady flow of
recreation seekers to prime sites from nearby towns, other
states, and even more distant regions. The consequences of this
situation are described by the ORRRC:[69]

> First, there is no logical place in the conventional
> government structure where responsibility to deal with
> this mercurial problem has been fixed. The problem
> certainly transcends the local community; but we find
> it also overflowing the region and the State, and at
> the very top, it spills out of the Nation. Then, the
> richly endowed community finds it increasingly difficult
> to have its exclusive claim to these riches recog-
> nized....What happens, in effect, is that the resource
> rich communities find themselves exporting tremendous
> volumes of free recreation services, frequently at a
> substantial social cost to themselves from the operation
> and maintenance of facilities and from the debasement
> of the recreation facilities to their own residents.
> One reaction has been to wall out the problem by restric-
> ting the use of the resource: Public beaches confined
> to the use of local residents, stream banks and wooded
> areas taken over by private "clubs." Carried to an
> illogical extreme, as such things sometimes are, the
> end result of this process is that a few have super-
> lative opportunities for outdoor recreation, while
> the great majority must compete for the services of a
> limited supply of mediocre-to-poor recreation resources.

It is not unreasonable to state that we are now approaching
(in many ways) that illogical extreme. Already we have noted
the "fencing out" tendency of local communities and private resi-

dents in Connecticut and even Maine, while fees and other restrictive measures to limit prime beaches to local residents
are commonplace on Cape Cod and in many other areas.

The particular *economic and political context* within which
local governmental units make decisions about shoreline use
can also lead to inefficient allocation on a broad scale.[70]
The political organization of the coastline is the historical
one of many small communities, each with control over a limited
segment of coastal property. The only political mechanisms
that are available to help correct private market deficiencies
in shoreline allocation are the local zoning and taxation policies of this fragmentation of local community control. We have
already seen how the uneven distribution of prime recreational
shoreline property places heavy demand pressures from the region
on specific communities, making their coastal property more
valuable than some neighboring towns not similarly "blessed" with
good beaches or scenic shore. Yet, in the absence of this value
being properly represented within the private market (where the
local community operates), local governmental units make decisions based on other considerations. This can best be illustrated by looking at the decision-making process involving a
specific coastal zone project, perhaps a power plant project.
It is important here to distinguish between two types of benefits
(or disbenefits) of such a project--*direct* and *indirect*. *Direct*
effects are those that accrue to the consumers or users of the
project, the user of the power supplied, the former bathers
on a closed beach, the swallowers of polluted air, the viewers
of marsh wildlife, etc. All of these effects are felt by the
local community and by the regional society in general. Yet
only those benefits (or disbenefits) that accrue to the local
populace enter into the decision. The community may be willing
to give up beach or bluff property to have a power plant which
will increase its tax base or bring a handsome profit to the
former owners of the site. However, this may not be an optimal
allocation of that resource on a regional basis. But the "votes"
of the region are not counted--only those of the local community

affect the decision!

We might ask why a community would be willing to give up this valuable property in such a way? The answer is that the local community within its particular economic and political context is also subject to the second type of benefits--*indirect* or *secondary* effects. These effects accrue to the suppliers of the resource that make the investment possible. For example, construction workers who build the plant may spend a substantial portion of their paychecks in the locale of the plants, certainly benefiting local merchants, doctors, and bar owners. These people, in turn, spend some of this money in the locale, and so on, in the traditional multiplier effect. Values that arise in this manner are also called *parochial* effects and include the effects on local payrolls, retail earnings, and the broadening of the tax base (usually a very powerful factor). For the local community, these benefits are very real; but, considering the regional economy as a whole, parochial benefits are *not* net benefits since the secondary benefits associated with one location will be about the same as those associated with an alternative site (barring large unemployment differentials). Thus, parochial benefits represent a transfer payment from one place in the economy to another, with no net benefit associated with the choice of site (even though there is a net benefit to the community chosen). Yet parochial benefits can be overwhelmingly impor- tant to political bodies representing the local community. As a result, a local community can rationally view a project in a very different manner from the regional economy as a whole. The region and the local community feel positive and negative direct effects such as the power generated or the beach lost-- the community alone feels the parochial effects such as a broadened tax base. These added benefits will persuade the community to act in its perceived self-interest and approve the power plant siting, with no consideration of the negative direct effect to the region as a whole. Such things have happened on the Maine coast where much of the loss of shoreline property came "with the encouragement of state and local agencies and officials

eager for new taxable property and the jobs that developments
generate."[71]   John McKee, the Bowdoin land-use expert, has said
"it is surprising how many people will sacrifice their coast.
They say, if it'll bring in the tax dollar, let's do it."[72]

Sometimes other forms of very localized political pressures
hinder effective planning for coastal land use and management.
A case in point is that of the tiny town of Harpswell on the
Maine coast. In 1969, a Planning Board was created to assist
the selectmen in considering some of the questions related to
the future growth of the town. By the 1970 town meeting, the
board had developed the preliminary plans for a three-year project and had obtained the needed appropriations from the local
voters. A chronology of subsequent events has been described in
a recent edition of the Maine Times.[73]

> With a complete map of the town and a detailed soil
> map, the Planning Board plunged into the nitty-gritty
> of formulating its first land use ordinance. The threat
> of unregulated subdivision seemed urgent. Out-of-town
> developers had purchased a 400-acre plot on Great Island,
> and some 3000 additional acres would soon be up for grabs.
> Less than one-tenth of the town's 24 square mile area
> had already been developed, and there were no local
> restrictions to inhibit irresponsible exploiters of the
> land.
>
> The Board also sensed some danger from within. Local
> developers had divided their land into undersized lots
> for trailers and small houses. Operating on shoestring
> budgets they built inadequate roads--impassable to school
> buses and fuel trucks in the spring--and petitioned
> the town to take over road maintenance and/or improve-
> ment.
>
> Slowly working their way through three rough drafts,
> the Board and its consultant, John Atwood, created an
> ordinance aimed at developers whose land use practices
> (insufficient soil surveys, inadequate sewage and water
> systems, narrow roads) would not be in the best inter-
> ests of the Town of Harpswell. Nine detailed sections
> dealt with the necessity for both a preliminary and
> final subdivision plan to be submitted to the Board;
> design standards for streets, sidewalks, lots, schools,
> etc.; performance guarantee; character of the develop-
> ment; variations and exceptions.
>
> The release of this proposed plan triggered a great deal of

political infighting. Some Harpswell citizens opposed any form
of land-use regulation and were vociferously against the plan
(and seemingly the word "planning" itself!).

> Some interpreted the ordinance to mean that they would
> have to demolish their own homes, remove lobster traps
> from their yards, receive permission to cut down a tree
> on their property. Others were moved to protest that
> it discriminated against the poor, the young, the large
> family. Discussion was propelled by emotional arguments
> which the Board was unable to direct into more reason-
> able channels.

The anti-planning group gained strength, including among its sup-
porters several local developers and contractors (who supplied
bus transportation for local voters to the town meetings). As
a result, the plan was defeated and the planning board was abolished at the next town meeting.

> Harpswell's future was on the line, and she stood
> defenseless before those who cared not for the common
> heritage of coastal land...with no planning board and
> no land use laws, Harpswell waits naked for the devel-
> opers' invasion.

> What happened in Harpswell...could have happened in
> any Maine town that has not yet confronted the ques-
> tion of its destiny.

The problems within the institutional environment are further complicated by the attitudes that some states have previously held in failing to regard the coastline as a separate
resource in need of regulation on the state level. For example,
in 1967 Maine citizens approved a four-million-dollar bond issue
for park and coastal acquisition, even though the legislature
insisted on a provision prohibiting the use of eminent domain
powers. Yet, as of mid-1970, "though prices in the meantime have
doubled and quadrupled, and tens of thousands of desirable acres
have changed from open space to luxury developments, the State
Parks and Recreation Department has spent only $567,000, less
than 12 percent of the money the voters authorized. And only
part of the purchases have been coastal property."[74] This, of
course, is only part of a larger overall problem with institutional involvement in coastal allocation. *In the absence of any*

long-range plans, local and state governments usually take an incremental approach to satisfying increasing demands for shoreline recreation (and most other things, for that matter). Most governmental units react only to short-range problems of supply and demand for shoreline facilities because of a lack of funds. This is understandable to the extent that states and local governments do not have the large amounts of money necessary to compete on the private market for all the coastal land that is needed, since there is no existing mechanism by which the values of the users of a public beach can be measured and translated into revenues. Thus, the only choice for government is to try to buy small stretches of shoreland when it is needed, to plan only for the demands of the next five or ten years. But, while this has been going on, potential sites have been privately bought and developed to the point where, as we have seen, practically nothing remains to be acquired.

3. Summary

Having recognized the high intrinsic value of our shoreline as a recreational resource in need of careful allocation, we have found the private market to be inadequate for the task of allocating this resource within the present socioeconomic and institutional environment. We conclude that market mechanisms will result in an allocation of the coastal zone which may be seriously inconsistent with the values of regional society. Standard market imperfections such as undervaluation of public goods and side effects work to price the general public out of the market for recreational areas without having a similar effect on private, commercial, and industrial development. In addition, the political organization controlling the use of shoreline land through local zoning and tax policies also contributes to a misallocation of coastal resources since, even if each community operates optimally within its own bounds, the total shoreline allocation will *not* be optimal, due to the lack of consideration of alternatives in which one community specializes in certain shoreline functions, while another specializes in some other activity. Local planning

may even lead to allocations that are *worse* than those of
an unrestricted market, since whenever a local board is faced
with a development proposal, its first thought is toward the
immediate secondary or parochial benefits of the project: the
effect on local payroll and retail earnings, broadening of the
tax base, etc. Yet these benefits are not *net* benefits, but
transfer payments from some other part of the regional economy.
In addition, the planning procedure of meeting increasing demands
on an incremental, piecemeal basis clearly wastes opportunities
for acquisition of valuable coastal acreage that is rapidly
bought and developed for private, commercial or industrial use.
The absence of any long-range planning on the part of state
and local governments has clearly contributed to the formation
of the crisis we face today.

The final question to be resolved is: given the inadequacies of the present system and the critical need for coastal
zone allocation consistent with the values of society, what are
the alternatives to the present allocative mechanisms? It is
clear that something must be done right away to satisfy the
demands of the immediate future; but there are also serious questions of long-range policy to be considered along with current
needs.

VII. A NEW FRAMEWORK FOR COASTAL ZONE MANAGEMENT

We have argued that the present allocative mechanisms of
the private marketplace and local political decision-making are
sorely deficient in their ability to respond to the needs and
demands of American society for shoreline recreational resources.
Hence, we must turn to a consideration of some new economic and
political framework that will correct these deficiencies. Immediate steps must be taken to formulate a policy that successfully comes to grips with the complex issues that are raised
when the present system is rejected. The purpose of this section is to focus attention on the political and economic ques-

tions that must be dealt with in the formulation of long-range
policy. In Section VIII we will make some specific suggestions
as to measures that might be effective in the short run.

If the allocation of the shoreline as a public good is to
to be handled in the public sector, then the first requisite is
the development of some alternative political framework within
which management of the resource can take place. We have seen
that the present framework of local political decision-making
is wholly inadequate, while no political mechanisms exist that
deal with our coast as a separate and unique entity. Yet clearly
the social costs and benefits of shoreline recreation go beyond
every municipal boundary and spill over from state to state.
It is clear that new institutional arrangements must be made
so that long-range, comprehensive planning policies can be formulated in the development of a recreational system and to determine
that allocation of coastal resources among multiple uses which
maximizes the benefit to society in general. What are these
new arrangements to be?

A major criterion that should be applied to new institutions
is that the political decision-making unit affecting any particular use of coastal resources must be sufficiently broad so
that the parochial benefits of a given development project are
not net benefits within the unit's jurisdictional boundaries.
We will recall that a coastal town may decide to zone its coastal
property for industry, generating (secondary) benefits for the
town, but not for the regional economy as a whole (e.g., if
the area is a valuable beach site). If a state or regional
body makes the decision, a broader consideration of benefits
to the regional populace should result. In this way, a greater
range of social costs and benefits can be weighed in the decision-
making process! The clear implication is that planning for
the use of coastal resources must be carried out at a more broadly-
based governmental level.

1.   Recent Legislative Activity

Much careful consideration has recently been given to this issue at the federal level, beginning with a report[75] to the President and the Congress in 1962 by the Outdoor Recreation Resources Review Commission (ORRRC). This study outlined the status of outdoor recreation in America, describing in depth the conditions of supply and demand as outlined in this article. To resolve the problems of shoreline recreation, the ORRRC called for the establishment of new guidelines for planning and policy and the design of new institutional *relationships* to manage the complex set of interdependences in a systematic way. These relationships would entail a redistribution of responsibility among governmental levels, with the *states* playing the pivotal role and the federal government taking on the responsibility of developing and maintaining the viewpoint and interests of the national system as a whole, while coordinating activities and *providing a mechanism for the resolution of conflicts between states.*

More recently, a number of other studies have made recommendations as to the proper political framework for sound coastal zone management. These include:

1) A report by the Commission on Marine Sciences, Engineering, and Resources.

2) The "National Estuarine Pollution Study," sponsored by the Water Pollution Control Administration of the Department of the Interior (November 1969).

3) The "Coastal Zone Management Conference," House of Representatives Subcommittee on Oceanography of the Committee on Merchant Marine and Fisheries (October 1969).

4) A Report to the Committee on Multiple Use of the Coastal Zone of the National Council on Marine Resources and Engineering Development on "Coordinating Governmental Coastal Activities" (September 1968).

1)   Report of the Commission on Marine Sciences, Engineering,
     and Resources

     This commission, headed by Dr. Julius A. Stratton, was
formed in 1966 by the Marine Resources and Development Act and
charged with the responsibility of formulating a program of
national action for the most effective use of our marine resources and a plan for governmental organization for the fulfillment of that program. The relevant recommendations of that commission were put forth in a statement by Dr. John A. Knauss,
former Chairman, Panel on Coastal Zone Management of the Commission:[76]

> A major conclusion of our Commission was that the
> primary problem in the coastal zone was a management
> problem with all the attendant problems that proper
> management implies. It is true that the Federal, State,
> and local governments share the responsibility to devel-
> op and manage the coastal zone. In reviewing the situa-
> tion, *we concluded that effective management to date
> has been thwarted by the variety of Government juris-
> dictions involved at all levels of Government, the low
> priority afforded to marine matters by State govern-
> ments, the diffusion of responsibility among state
> agencies to develop and implement long range plans*...the
> Commission was of the opinion that the *states* must be
> the focus for responsibility and action in the coastal
> zone. We believe an agency of the State is needed with
> sufficient planning and regulatory authority to manage
> coastal areas effectively and to resolve problems of
> competing uses.
>
> We recommend that a Coastal Management Act be enacted
> which will provide policy objectives for the coastal
> zone and authorize Federal grants-in-aid to facilitate
> the establishment of State coastal zone authorities em-
> powered to manage the coastal waters and adjacent land.
> (Emphasis added)

2)   Department of Interior Report--The National Estuarine
     Pollution Study[77]

     The recommendations and proposed program outlined in this
report put forth the policy objectives for a comprehensive
National Program for Coastal Zone Management and spelled out the
suggested reponsibilities and roles of the Federal, State and
local governments within such a program.

What is proposed is a program that recognizes the
primary responsibilities of the States in a management
program for their estuarine and coastal areas, and on
the Federal side provides for the coordination of
Federal activities in these areas and for assistance
to the States in their management activities.

Any comprehensive national program for the estuarine
and coastal zone must provide flexibility in many ways
to fit regional and local conditions and situations,
but regardless of variables *it must establish a guiding
policy and a set of objectives*. Regardless of variables,
in order to be effective the program must provide for:
1) Planning and implementation; 2) active administration,
coastal coordination, and financing; 3) the development
of the knowledge and data necessary as a basis for all
action.

The recommended National Policy will put in effect
a comprehensive national program for the effective
management, beneficial use, protection and development
of the estuarine and coastal zone of the Nation involving Federal, State, and local governments, and public
and private interests in an appropriate manner. It
will permit the optimum use of this vital resource by
recognizing the existence of competing uses and accom-
modating them through appropriate management and,
further, conserve these resources in such a manner as
to keep open the options for various uses in the future
and not foreclose them. This management system will
recognize the primary and constitutional role of the
States in managing their resources as well as the role
of the Federal Government in protecting the wider national interest. The principal goal of the national
program is the use of the estuarine and coastal zone
for as many beneficial purposes as possible and, where
some uses are precluded, to achieve that mix of uses
which society, based on both short- and long-range considerations, deems most beneficial. (Emphasis added)

3)   Coastal Zone Management Conference[78]

These hearings brought together a wide range of parties involved in the problems of coastal zone management. The tone of
the conference can best be illustrated by excerpts from some of
the testimony given therein:

Statement of Dr. Samuel A. Lawrence, Former Executive
Director, Commission on Marine Sciences, Engineering
and Resources

...We need to establish a firmer legal framework for

ownership and use of coastal and offshore lands. Above
all, the commission concluded, the pressures for multiple use of these limited coastlands require an organized approach in order to coordinate the separate plans
and activities of Federal, State, and local government
agencies and of private persons and corporations.

Statement of John R. Quarles, Assistant to Under-Secretary
for Environmental Planning, Department of the Interior

There appears to be developing something approaching a
consensus that responsibility should be vested primarily
in the State government and exercised at the State level.

...I don't believe anyone who has seriously focused on
the problem thinks that the Federal government can,
from Washington and from the Federal level, devise
management plans which properly would anticipate the use
that each acre of land should be devoted to over the
years ahead, so the Federal government needs to be ruled
out as being the primary responsible agency in management of coastal areas.

...The localities, I would suggest, are not suitable
for exercising these functions. It has been fairly
widely recognized that localities suffer from deficiencies of not having strong staffs, skilled people to deal
with some complex problems. Also, of course, they are
extremely concerned with development of their individual
tax bases of assessable property within the town limits.

These considerations, however, I would suggest, overlook
the principal difficulty with leaving responsibility at
the local level, which is that good planning from this
time forth needs to encompass a range of vision beyond
town limits....Development cannot be done well on a
local basis....I think that, regarding this problem on
a national level, serious consideration must be given
to whether we can continue to allow areas which can be
seen as needed to meet other needs to be used for residential development.

4) Report to Committee on Multiple Use of the Coastal Zone of
    the National Council on Marine Resources and Engineering
    Development on "Co-Ordinating Governmental Coastal
    Activities"[79]

The primary aim of this study can be described as follows:

- to recommend means of coordinating governmental agencies
  acting in the coastal zone;

- to identify gaps, overlaps and inadequacies of coordination in the activities of federal agencies in the coastal
  zone and to recommend appropriate solutions;

- to identify the need for improving federal-state relationships in the coastal zone and to recommend appropriate solutions.

The report identifies four basic uses of the coastal zone:
enjoyment, transportation, national defense, and *land use.*
Some of the recommendations and conclusions regarding land use
and enjoyment are directly pertinent to this study. One such
recommendation urged that the Department of the Interior lead a
multi-agency study to *propose national objectives and goals
for enjoyment of the coastal zone.* This study would "address
such matters as 1) the relative roles and values of low-density
enjoyment, such as preservation, conservation, hiking and hunting
vis-à-vis high-density enjoyment, such as bathing beaches, marinas, athletic facilities and entertainment; 2) future recreational requirements, their types, quantity, and distribution; and
3) rational, measurable objectives, related to economic benefits
achieved and foregone, to help fill the partial void now facing
federal agencies when tradeoff decisions must be made between
quantifiable economic effects and many as yet unquantified enjoyment values."

The report goes on to conclude that considerable effective
federal-state coordination can be obtained "through improved,
tempered use of such means as the normal course of business:
informational services, mutual assistance, grants, subsidies
and regulations. Where interstate conflicts arise that could
not be handled by existing institutions (river basin commission,
etc.), new institutional arrangements should be created."

As a result of the conclusions and recommendations set forth
in these and other studies on coastal zone management, considerable attention has recently been devoted to the forming of new
legislative proposals at the federal level. Among them are:

1)   A bill (S.2802) to assist the States in establishing
     coastal zone management programs--introduced by
     Senator Warren Magnuson (D.-Wash.), August 8, 1969

2)   A bill (H.R.14730) to provide for the effective
     management of the Nation's coastal and estuarine
     areas--introduced by Representative Alton Lennon
     (D.-N.C.), November 6, 1969

3)   A bill (S.3183; H.R.14845) to provide for the estab-
     lishment of a national policy and comprehensive na-
     tional program for the management, beneficial use,
     protection, and development of the land and water
     resources of the Nation's estuarine and coastal zone

     - introduced in the House of Representatives by
       Representative Fallon, November 18, 1969

     - introduced in the Senate by Senator Boggs,
       November 25, 1969

It would be useful to examine the provisions included in these
bills so that we can compare them to the emerging concepts that
make up the new political framework for coastal zone management.

1)   S.2802

     This bill, recognizing the harmful side effects of unplanned
and poorly-planned development of coastal resources, declares
that the policy of Congress is to "preserve, protect, develop,
and where possible restore the resources of the Nation's coastal
zone...through comprehensive and coordinated long-range planning
and management designed to produce the *maximum benefit for society* from such coastal areas." To facilitate such planning at
the State level, the National Council on Marine Resources and
Engineering Development may award grants-in-aid (or underwrite
bond issues or loans) to *coastal authorities* (designated by
the Governor of a Coastal State through legislative or other
processes) to assist them in developing a *long-range master plan*
for the coastal zone and in implementing a *development program*
based on such a plan. To secure the Council's approval the State

plan must:

- set forth desired goals and standards;

- promote the balanced development of natural, commercial, industrial, recreational, and esthetic resources and to accommodate a wide variety of beneficial uses;

- estimate future population and the needs of the above competing uses for coastal land;

- include diagrams for the most efficient, beneficial, and liveable interrelationship of these uses;

- gather information on the existing land-use regulations and consult with various governmental bodies whose jurisdiction extends over territory located in the coastal zone (local, regional, port, intrastate, and Federal authorities).

In addition, the bill provides authority for the development of the coastal zone in accordance with the master plan through the use of land-use and zoning regulations, acquisition of lands through condemnation or other means, and the issuance of bonds. Also, the coastal authority has the authority to review all State and local projects and to reject developments that do not comply with the principles and standards set forth in the master plan.

2)   H.R.14730

Finding that the rapidly intensifying use of coastal and estuarine areas has outrun the capabilities of Federal, State, and local machinery to plan their orderly development and to resolve conflicts, this bill declares to be the policy of Congress to foster the effective utilization of coastal and estuarine areas through assistance to coastal states in the development of a management system permitting conscious and informed choices among development alternatives. This Act would empower the Administrator of the National Oceanic and Atmospheric Agency to make grants (or underwrite bonds and loans) to state coastal

authorities provided that the authority submit and obtain approval of a long-range planning proposal that must incorporate a
number of particular considerations as outlined in the bill.
These include:

- identification of the coastal areas requiring concerned
  attention and development of a plan for their most
  effective utilization;

- provision of machinery for the resolution of conflicts
  arising from multiple use;

- provision for necessary enforcement powers through zoning,
  permits, licenses, easements, acquisition or other means
  to assure compliance with plans and resolve conflicts
  in uses;

- provision for coordination with local, State, and Federal
  agencies, research institutions, private organizations,
  and other groups as appropriate to provide a focus for
  effective management;

- fosters the widest possible variety of beneficial uses
  to maximize social return, achieving a balance between
  the need for conservation and for economic development;

- takes into account the rights and interests of other
  States and respects Federal rights.

3)   S.3183; H.R.14845

This legislation, submitted to the Congress by former
Secretary of the Interior Walter J. Hickel, is based on the
findings of the National Estuarine Pollution Study and an interdepartmental Coastal Zone Task Force chaired by Under Secretary
of the Interior Russell E. Train. Under the provisions of the
bill, the Secretary is authorized to make grants to any coastal
state for the purpose of assisting in the development of a comprehensive management program for the land and water resources
of the coastal zone. In order to qualify for such grants, the
coastal state must demonstrate to the Secretary compliance with

the following requirements:

- the coastal state must be organized to implement a
  management plan;

- the agency or agencies responsible for implementation
  must have the regulatory powers necessary to implement
  the plan, i.e., permit authority, authority to acquire
  interests in land through eminent domain and zoning,
  authority to require conformity of local zoning to
  the State plan;

- the coastal state has developed and adopted a management plan for its coastal zone;

- the plan must include identification and recognition
  of national, state, and local interests in the preservation, use, and development of the coastal zone;

- the plan must identify and describe means by which the
  management proposal will be coordinated with interstate and regional planning;

- the plan must be developed in cooperation with relevant
  Federal, State, and local governments, and all other
  interests;

- the plan must develop a feasible land- and water-use
  plan, reasonably reflecting the needs of industry,
  transportation, recreation, fisheries, wildlife, natural
  area protection, residential development and other public
  and private needs, both in the short and the long term.

The bill makes additional provisions for interagency coordination and cooperation on the Federal level.

> The Secretary shall not approve the plan submitted by
> the State...until he has solicited the views of Federal
> agencies principally affected by such plan or his evi-
> dence that such views were provided the State in the
> development of the plan. In case of serious disagree-
> ment between any Federal agency and the State in the
> development of the plan, the Secretary shall seek to
> mediate the differences....Federal agencies shall not
> approve proposed projects that are inconsistent with

the plan without making investigation and finding that
the proposal is, on balance, sound. The Secretary shall
be advised by the heads of other agencies of such problems and be provided an opportunity to participate in
any investigation.

2. Establishing a New Political Framework

Based on the analysis in this article and the conclusions
and recommendations of the aforementioned studies, we are now in
a position to outline some of the considerations that would go
into the legislative formulation of a new political framework
for coastal zone management. It appears clear to us that the
destiny of shoreline resources should be removed from the hands
of local decision-making and entrusted to a broader-based governmental unit.

Many of our present-day problems, such as air and water
pollution, electric power production, and land use, are inherently *regional* in nature and could seemingly be handled most
efficiently by *regional* governmental bodies. But if we are to
assume that it is desirable to work within the existing governmental structure, then it seems that the *states* should play
the primary role in coastal zone management. We must be careful, however, to realize that even the state may not be broadly
based enough to handle many coastal land-use problems. We have
noted how the trend toward increasing mobility and the uneven
distribution of suitable coastal opportunities has made the
problems of shoreline recreation ignore all state and local
boundaries, especially in New England where coastal facilities
are often within a two-hour drive from many parts of the region.
The problems of inefficient allocation, which arise because decisions are based on considerations of secondary benefits, are
not restricted to the local communities. This could happen at
the interstate level, especially when there is a large discrep-
ancy in the economic posture of two nearby states. For example,
Massachusetts is a well-developed and economically healthy State
with a large population, while Maine is economically depressed
and low in population. Hence, these States might take a differ-

ent orientation toward the development of the Maine coast. The
State of Maine might welcome oil refineries and industrial complexes as a stimulus to the State economy, while the residents
of Massachusetts value the coast as a unique recreational opportunity, especially since Massachusetts' shoreline facilities
are already used to capacity. But the benefits and disbenefits
to the regional society outside of Maine's boundaries will *not*
be included in the determination of the costs and benefits of
particular development projects. Hence, the state will make
decisions based in part on the parochial effects to the state economy; yet this may constitute an inefficient use of the resource.
Any new political framework for coastal resource allocation
that has the state as the focal point for management must devote
careful attention to problems of this sort.

A second major consideration pertinent to the management
of the coastal zone in the public sector is the question of *how*
decisions are to be made regarding shoreline resource allocation between competing uses. If we conclude that the allocative
mechanisms of the private market are inadequate, then the State
and Federal management authorities must have some alternative
means for determining what is an efficient allocation of the
shoreline. This must necessarily involve the determination
and articulation of the *public interest*. In the private market,
goods have a mechanism (price-profit system) whereby the demands
of individuals can be felt; when the aggregate of individual
demands is high enough, private producers will attempt to satisfy those demands. Thus, many individual preferences can be
satisfied, since each individual's "vote" (in dollars spent)
goes relatively far in determining the supply. Whenever enough
individuals want something at a price, there is an incentive
for someone to produce it at a profit. Public goods differ
in that private markets fail to respond to the entire range
of individual demands, giving rise to a need for collective
action. The question is, how can individual preferences for
these goods be summed to determine if the aggregate benefit
is sufficient to justify the total cost? This is a central

question in the area of welfare economics, and the resolution
of the issues involved must play an important role at the Federal
and State levels in the formulation of management policy concerning
coastal land use.

A number of theories have been set forth involving this
crucial determination of the public interest. The point of
view of an *aggregated social welfare function* holds that society
maintains a hierarchy of priorities based on collective values,
inviting a search for the articulation of these priorities.
A fundamental question to be dealt with in this regard is:
Are these social priorities effectively articulated through
the democratic political process as it now exists so that decision-
makers are adequately equipped to act in the public interest?
Another point of view is that of *willingness to pay*, which holds
that the maximum amount of resources that consumers are willing
to pay for a good is a good measure of its value. This can
be expressed as a willingness to pay additional taxes, user
fees, and other charges, to give up the consumption of certain
goods, or to pay a higher price for other goods. The primary
objection to this scheme is based on the difficulty in measuring
the willingness to pay for public goods that are not "unitized"
and whose benefits to an individual are hard to determine.
Cost-benefit analysis uses willingness to pay and appears to
have the potential for effective simulation of the working of
a properly functioning market in the allocation of some public
resources on a project-by-project basis. Such an analysis has
been demonstrated and recommended in a report[80] to the Water
Resources Council by the Special Task Force on Evaluation Procedures
as a way to improve the policies and procedures followed by
Federal agencies in the formulation and evaluation of projects
for the use and development of water and related land resources.
More recently, a report[81] on economic factors in the development
of a coastal zone has described a preliminary effort at the
development and application of such analysis to particular coastal
zone development proposals:

The basic premise of this report is that economics
in a sense wide enough to cover all significantly
important values, both market and nonmarket, can be
usefully applied to coastal zone allocation, that is,
to the problem of determining the mix of uses of a
particular coastal zone which is most consistent with
the values of the economy which uses that coastal zone.

Given the inefficiency of the private market with
respect to the coastal zone and the inefficiency of
local control, the only feasible alternative appears
to be control at the state level with some federal
influence to prevent secondary benefits from being used
against an entire state. We strongly support the
Stratton Commission's recommendations concerning the
establishment of state coastal zone management
authorities.

However, the establishment of such bodies implies some
rather heavy responsibilities. *Once the discipline of
the private market is abandoned, coastal zone analysis
requires conscious economic analysis, for it is not
enough to show that the present system is seriously
nonoptimal. One must also argue that the proposed
changes in the allocation process will result in
coastal zone usage which is more consistent with the
economy's values than the old.*

Insofar as coastal zone allocation can be regarded on
a project-by-project basis, the methodology for implementing this conscious economics is cost-benefit analysis. Unfortunately, the present state of the art with
respect to cost-benefit analysis and the coastal zone
leaves much to be desired and, until a state coastal
zone authority can reliably determine the use of the
coastal zone most consistent with people's values, it
cannot promise to do much better than the private market or local political entities.

Another problem with locational cost-benefit analysis
is that, if performed too narrowly, seriously inefficient suboptimization can occur. The problem is to
approach coastline allocation comprehensively while,
at the same time, retaining analytical feasibility.
Given the compromises that must necessarily occur, the
results of cost-benefit analysis must be used with
some judgment. (Emphasis added)

While there seem to be no clear-cut indications that any
method of determining the public interest is superior to the
others, this is no excuse for inaction--attempts must be made
to determine the public interest. Perhaps the answer lies in

some combination of the viewpoints of *representative political consensus* (based on overall social priorities) and *cost-benefit analysis* (based on willingness to pay) as effective measures of the public interest. The important point is that some determination must be made, both at the Federal and the State levels, before we can claim that the new framework for coastal zone allocation is *better* than the old one of the private market and local political decision-making.

Having warned of the dangers of interstate side effects and the need to determine carefully the public interest, let us now attempt to outline the roles of the State and Federal Governments in a sound coastal land-use management system.

The State Role

The role of the states in coastal zone management is recommended to be as follows:

1)   To assume *primary responsibility* for the planning and implementation of a comprehensive coastal land-use management plan to bring about effective utilization of shoreline resources most consistent with the values and interests of national, regional, state, and local society.

2)   To establish some form of coastal zone authority empowered to develop a master plan for coastal land and water management and to implement this plan through the use of any legal means, such as zoning, permits, licenses, eminent domain, easements, acquisition, issuance of bonds, etc.

3)   To develop a master plan that has the following characteristics:

    - sets forth desired goals and objectives consistent with the values of society at all levels (local, state, regional, national);

    - *establishes guidelines for the determination of the public interest consistent with similar efforts at the Federal level;*

- provides a mechanism by which decisions can be
made regarding the efficient allocation of
coastal resources among the competing uses and
needs of industry, recreation, commerce, transportation, residential development, wildlife and
natural area protection, etc., *based on the established goals and guidelines for the determination
of the public interest;*

- provides for coordination and cooperation in the
development of the plan with local, state, regional
and federal agencies and any other public or private organizations with a vested interest in
coastal land-use management, and is consistent
with planning efforts at all the various governmental levels;

- provides up-to-date inventories and evaluations
of the status of shoreline resources within the
State's jurisdiction, including the accessibility
and suitability of beach, marsh, and bluff areas
for various uses.

The Federal Role

The role of the federal government in land-use management
in the coastal zone is recommended to be as follows:

1)   To provide the overall political framework within
      which the planning efforts of the individual coastal
      states and the various federal agencies can be *coor-
      dinated* in the development of an efficient land-use
      program that is compatible with not only statewide,
      but also regional and national interests and values.

This first function of the federal government in land-use management in the coastal zone entails substantial responsibilities.
These responsibilities come directly from the need to coordinate
the planning activities at the state levels and to resolve serious conflicts that might lead to a grossly inefficient allocation
of resources due to the existence of statewide secondary (parochial) benefits. The key concept here is *coordination;* since

many of the problems of coastal land-use management are inherently *regional* in nature, it is not enough to stop at the establishment of state coastal authorities in the formulation of a political framework. While such authorities seem to provide an
effective means to overcome the problems attendant upon *local*
decision-making in the presence of secondary benefits, they do
nothing to solve the problems of *interstate* conflicts of interest
that come about for the same basic reasons. Indeed, the issue
is the same but occurs at a different governmental level! Yet
there are *no* political mechanisms to resolve such conflicts at
the regional level, where these problems might best be handled.
This underscores the necessity of the federal government's taking
an *active* role in coordinating the planning efforts of the states
and filling the void created by the absence of regional decision-
making units. This might be effectively realized through the
creation of a national land-use agency or commission, subdivided
into groups that are to take a regional orientation toward the
coordination of state land-use management programs. The charac-
teristic activities of such an agency would include the following:

a)   provision of the financial and informational
     basis of support for the planning and implementation
     of state and regional land-use plans, based on a
     review and approval of such plans;

b)   encouragement of the cooperation of neighboring
     states in the development of a regionwide land-use
     master plan, possibly through the formation of
     regional land-use authorities;

c)   coordination of the activities of all the federal
     agencies in relation to land-use management and
     development of mechanisms to resolve interagency and
     federal-state conflicts.

Up to this point, the formation of a new political framework
has dealt primarily with the problem of more effective government
coordination of activities with regard to coastal zone management. We have given considerable attention to the need for a

more broadly-based governmental body to manage coastal land resources and to avoid the gross inefficiencies that have come
about due to the uncoordinated activities of local political
decision-makers. But again we must remember that this is only
one side of the story; we have also decided that the private
market is unsatisfactory in the allocation of scarce shoreline
resources. This presents us with the difficult circumstance of
having to make decisions based on tradeoffs between some very
quantifiable benefits and other inherently nonquantifiable
values. *The fact that we have rejected the discipline of the
private market in its present form does not mean that the circumstances that led to its failure as an allocative mechanism must
no longer be confronted.* The same kinds of decisions remain
to be made! Indeed, this is an indication that we must redouble
our efforts concerning the identification and articulation of
the values and interests of society, since we no longer can
rely on the relatively automatic workings of the price system,
which has performed this function for us in the past. *We cannot
assume that the problems of inefficient coastal zone management
can be solved by political reorganization alone.* Poor decisions
have been made in the past by local governmental units and by
the economic system itself. Correcting the political problem
is only half the solution; we must now face the issue of *how
to make decisions in the public sector that are consistent with
the values of society.* This is, as we have seen, no easy task.
It requires concerted effort at both the state and federal level.
This points to the second major function of the federal government
in coastal zone management, to be carried out within the coordinating
framework outlined above:

> 2)   To establish uniform *goals* and *objectives* that are
>   an effective articulation of the values of society
>   at all levels, and to set forth consistent *guide-
>   lines* for the state to follow in the formulation
>   of coastal land-use management programs that will
>   lead to the achievement of these objectives.

This implies that it is not sufficient to assume that the states
on an individual basis can provide mechanisms through which

decisions can be made as to the most beneficial allocation of a
particular coastal resource. The states must have the capacity
to make decisions based not only on intrastate values, but also
on regional and national interests. Yet the orientations of
different states towards what is really in the national and regional interest are likely to be widely divergent and heavily
weighted by the particular values of the people of each state.
The other implication is that the states need help in determining
how to measure and weigh the values of the people within their
own jurisdiction. This gets back to the ideas of representative
political consensus and cost-benefit analysis as effective arti-
culations of the public interest. The failure of any one state
to handle this crucial issue in a successful way would necessarily have a deleterious effect on an entire region due to the
intraregional nature of land-use management problems. To assist
the states in activities of this sort, the proposed national
land-use agency should support in-depth investigations into a
number of substantive issues of national concern in the area of
coastal land-use management with the purpose of establishing
guidelines in the following areas:

a)   how to make decisions in the public sector that
      involve tradeoffs between quantifiable economic
      benefits and nonquantifiable economic values;

b)   how to deal with circumstances in which tradeoffs
      between basic rights in a free democratic society
      seem unavoidable, e.g., the right to own, control,
      and develop personal property versus the right to
      swim at an ocean beach or explore a rocky bluff;

c)   how to eliminate as many conflicts in land use as
      possible through the implementation of innovative
      technology, e.g., by encouraging the siting of elec-
      tric power plants or other industrial complexes at
      offshore locations rather than in ecologically
      fragile estuarine zones (see Reference 58);

d)   how to include both the quantifiable and nonquanti-
      fiable values of regional and national society in
      the decision-making process at the state level;

e)   how a *regionwide* plan, once determined, might
      effectively be implemented using the legal tools at

the disposal of each individual state, e.g., effluent discharge fees, etc.

This completes the outline of a new political framework for
the allocation of shoreline resources and the management of land
use in the coastal zone. Let me now turn to a comparison of
these concepts with those set forth in the coastal zone bills
cited previously.

It seems clear that, while the various coastal zone management bills now under consideration have established the role of
the states in a substantive way, they have at the same time ignored the most crucial recommendations set forth in every study
as to the role that the federal government must play in the
overall management system. Each bill calls for state coastal
authorities to develop plans that set forth objectives consistent with regional and national interests--yet none provides for
the establishment of uniform guidelines for the states to follow
in the determination of these interests. Each bill calls for
the states to provide a mechanism for the resolution of conflicts, fostering the widest variety of beneficial uses to maximize social return--yet none suggests the mechanism by which the
needs and values of neighboring states can be effectively included in the tradeoff analysis. In addition, none of the bills
makes provision for the establishment of national policy objectives and guidelines for planning by which the plans of the
various coastal states can be coordinated. Nor is there any indication of how the administrator at the federal level is to go
about determining whether or not each individual state's master
plan is consistent with the national interest. Only one bill
suggests a mechanism for the resolution of federal-state conflicts, while none tackles the crucial issue of inefficient allocation due to secondary effects between states. While all these
bills seem to effectively spell out the roles of the states in
coastal zone management and establish the financial and informational bases of support for such efforts, they are seriously
deficient in not providing for the strong federal involvement

that is necessary for two important reasons: 1) to establish
substantive policy objectives and guidelines for effective *coordi-*
nation, on a *regional* basis, of the separate activities of the
individual states; and 2) to take the lead in tackling the difficult issue of *how to make decisions* (at state, regional and
federal levels) based on tradeoffs between measurable and nonmeasu-
rable benefits and costs to society at all levels. Unless this
involvement is provided for at the federal level, any political
reorganization that relies on the primary role of the states
attacks only half the problem. Thus, we would have to be prepared
to accept, at best, halfway solutions. There is doubt in my
mind that this would be any improvement over the situation as
it exists today, inefficient as it certainly is. There is real
danger here: we run the risk, for all our well-intentioned
efforts, of creating more serious problems than those we are
striving to solve!

VIII. NEW ENGLAND SHORELINE RECREATION IN THE SHORT RUN

     Clearly, there is a pressing need for the formulation of
long-range policy with regard to our shoreline resources. Such
a policy might well entail some radical departure from the cur-
rently-accepted allocative mechanisms of the private market and
local decision-making. All indications are that, without such
changes, there is *no way* to provide adequate public recreational
opportunities at our coastal shores for the hordes of people who
will need and demand such opportunities in the future. But the
only way to avert disaster until such a policy is formulated is
to *buy time* with the traditional procedures of short-term planning. Such procedures have helped to get us into this mess, and
it seems ironic that they should serve to help us correct the
problems. But we must be wary that the problems get worse at
an increasing rate; hence, the time that can be bought with each
incremental measure gets shorter. Any recreation planner will
attest to this--as new beaches open, they are soon used extensively and frequently to capacity, depending on the location.
Also, it is clear that we are almost without room for further

expansion in terms of land acquisition. Therefore, one thing
must be done in the short run: *in the face of spiraling prices
for shrinking amounts of available coastal land, state and federal authorities must take immediate advantage of current opportunities for land acquisition, using all the legal tools available to them to preserve more shoreland for public recreational
activity.* In New England, there are still a few areas that
could be developed to expand the recreational opportunities
available to the people of the region. We shall first outline
the supply status of the shoreline recreational resources in
New England and then focus on two important locations where
planned development is necessary and desirable--Cape Cod and
the Boston Metropolitan region.

1. Shoreline Recreation Resources of New England

The distribution of shoreline resources in New England exhibits much the same pattern as in the rest of the country. In
every state only a small portion of the total recreational shoreline is publicly owned. Table 3.5 gives a state-by-state breakdown of the distribution of New England recreational coastline
by type of shore, ownership, and development status. The New
England shoreline supply has been extensively documented by the
ORRRC in its Study Report No. 4, "Shoreline Recreation Resources
of the U.S."

Maine

The status of recreation shoreline in Maine was extensively
discussed in Section II of the chapter and will not be elaborated
upon here.

New Hampshire

The ORRRC report of 1962 describes the New Hampshire shoreline as a succession of sand beaches separated by ledges or headlands of rock. The beaches are narrow and relatively steep, the
sand supply is limited, and erosion is a major problem (due to
these limited supplies) at Hampton and Rye beaches. Of all New
England shorelines, New Hampshire's is the smallest and most

Shoreline Type	Me.	N.H.	Mass.	R.I.	Conn.
Beach	23	7	240	39	72
Bluff	2,510	9	288	145	61
Marsh	69	9	121	4	29
Total	2,612	25	649	188	162
Ownership					
Public--Recreation	34	3	12	8	9
Public--Restricted	-	-	6	10	-
Private	2,578	22	631	170	153
Development Status	Low but rising rapidly	Very high	High	High	High

Source: U.S. Department of the Interior, Bureau of Outdoor
Recreation, "Shoreline Recreation Resources of
the United States," ORRRC Study Report No. 4,
(1962), p. 12.

Table 3.5   New England Recreational Shoreline (miles)

highly developed. Commercial, resort, and private activities
claim 22 of the 25 miles of recreational shoreline, with the
remaining three miles used as a public recreational area. The
shoreline is best suited for swimming and fishing, while the intensity of development and the absence of bluff shore makes
hunting, camping, hiking, or scenic activities not feasible.
New Hampshire has no pollution affecting coastal recreation.

In 1963, New Hampshire began a State Planning Project that
called for inventories of recreation facilities and a recreation
plan. This plan places emphasis on development to provide facilities to serve tourists, and stresses the acquisition of coastal
marshland. The 1970-1971 budget calls for $50,000 of state
funding for development construction at the 50-acre Hampton Beach
State Park. In 1976-1977, $20,000 (one-half to be federal fund-
ing) is to be used for landscaping 30 acres of filled land now
barren and unused.

Hampton Harbor will be further developed in 1972-1973. After a survey project, $150,000 (one-half federal) will be spent on development. There is erosion at the access to the harbor, which is in need of stabilization for parking and boat-launching facilities.

At Fort Dearborn in Rye, money has been allocated for the acquisition of 200 acres of marshland adjacent to potential park property to protect it from intrusion. In addition, $460,000 has been allocated for surveys and construction in 1974-1975. Present public use of 136 acres here is nonexistent although the state has owned the land for 14 years. The plan is to determine a specific use for the land and develop the site for future recreation.

In New Castle at Fort Constitution, two acres, now unfit for use, will be reconstructed. Also, an additional 125 boat slips will be built at Rye Harbor in 1972-1973.

Since the New Hampshire beach facilities are used to near capacity, future emphasis should be placed on the development of Portsmouth Harbor and the Piscataqua River Basin for marine and docking facilities for pleasure craft. Pollution control in this area is a crucial adjunct to any development plan.

Connecticut

The ORRRC report describes the 162 miles of Connecticut shoreline as extremely irregular, with many bays, coves, promontories, beaches, and rock exposures along the Long Island Sound shore. The nature of the coast is quite varied, with 72 miles of beach, 61 miles of bluff, and 29 miles of marsh, frequently located behind barrier beaches. The entire shore is subject to erosion of approximately one foot per year resulting from local wave action and storm damage. Although the shore is extensively modified by seawalls and other protective structures, some facilities developed forty years ago 50 to 100 feet back from high water have either been destroyed or have little beach area left. This problem is economically serious and is intensified by occa-

sional (1 per 15 years) severe damage due to catastrophic hurri-
canes.

While the shore is highly developed for commercial and private recreational usage, only nine miles are in public control
for recreation. It is not uncommon for public area usage to be
controlled by preferential resident admission, parking restrictions, and other regulations since many residents feel the need
to preserve the areas for the local populace in the face of
potential overcrowding from nearby New York residents.

The area is suitable for swimming, boating, sailing, and
other water sports, although pollution is a local problem in
several areas. Due to the high level of shore development,
hunting, camping, hiking, and scenic activities are not feasible.

Rhode Island

Rhode Island has 188 miles of recreational shoreline with
only eight miles developed for public recreation. The ORRRC
report describes the shoreline outside Narragansett Bay as a combination of rocky, low-bluff-type with isolated headlands and
extensive sand beaches. Inside Narragansett Bay the shore is
almost everywhere a low bluff fronted by a very narrow beach
strip of sand and gravel. Erosion problems are moderate and the
shoreline is relatively free of protective structures. Pollution
has closed many areas in the bay to swimming and the taking of
shellfish; raw sewage discharges have polluted the waters of
Mount Hope Bay in Bristol, Apponaug Cove in Warwick, and Jamestown;
and many other salt water areas are considered unsuitable for
swimming. Even in some areas with adequate treatment it is
unsafe to swim because of lapses in operations.

The Narragansett Bay shoreline is most highly developed for
private recreation use, while the open coastline is moderately
developed. Here, the unique combination of large ponds (behind
the sand beaches) and ocean beaches on the open coast has not
been exploited. In 1962, approximately 200 square miles of good
land were available in shore communities and 50 miles of beach were

practically unused.

The Rhode Island shoreline is well suited for swimming, sailing, boating, fishing, and other water sports, with some opportunities for hiking and camping.

Massachusetts

The Massachusetts shoreline, except for the Cape Cod region, is generally a rocky, low-bluff type of coast with numerous sand and gravel beaches and tidal flats. The Cape Cod shore is mostly marsh and beach, while the outer face of the cape is a continuous sand beach backed by high dunes. There are numerous small harbors and shelter areas along the entire coast, many of which are centers for sailing and boating. The shore is best suited for sailing, boating, fishing, and swimming (in certain areas). Hunting and camping activities are limited. Just about all of the most desirable shoreline has been developed for private use. Of the 631 miles of recreational shore, only 12 miles are publicly owned for recreation. Further restrictions to public use are caused by severe pollution in Boston Harbor, where two beaches have been closed to the public (Tenean in Dorchester and Constitution Beach in East Boston); one is closed periodically (Wollaston), and many islands with great recreational potential cannot be used at all due to the effects of sewage outfall and polluted tributaries.

As might be expected, the greatest recreational demands are home-based and centered in the Boston metropolitan region, with 70 percent of the participation occurring within an eight-hour time span. Cape Cod and the islands have the greatest intensity of use for recreation away from home. The land in the Boston region, although developed for high-density use, is used to capacity each summer. There is a need for more high-density development in this region. The City of Boston owns many of the islands in the harbor which could be put to good recreational use if the pollution problem could be solved. This is also true of the banks of the Charles River.

A Calculation

This overview poses serious questions of planning for the
shoreline recreational needs of future New England society. To
illustrate the severe nature of these problems we can make some
rough calculations to get an idea of how well the present supply can satisfy future demands. The 1965 National Survey of
Outdoor Recreation found that New Englanders lead the nation in
per capita participation in ocean swimming at 3.11 days per year.
Using this number (assuming all the swimming takes place in June,
July, August) we can show that, for a New England population of
11.5 million, approximately 400,000 persons use ocean beaches
each day during the summer. On the supply side, there are 66
miles of publicly-owned recreational shoreline, which we will
assume to be primarily useful beach coastline (this is necessarily conservative since some of this land, e.g., in Maine, is
bluff shoreline and cannot support the same density of users
as a beach). If we assume an average beach width of 50 feet
and apply the criterion of a minimum of 50 square feet per person (this is also conservative since a number of city and county
planning commissions have standards that call for 75 to 150
square feet of beach per person), then one mile of beach can
support approximately 5,500 people per day. Hence, simple extrapolation indicates that the total capacity of New England's total
public beach system is approximately 363,000 users per day. Thus,
the carrying capacity (363,000) of New England's public shoreline
seems to be in the general vicinity of present demands (400,000)
placed on it for ocean swimming activities.

While the overall conclusion seems inescapable, many of the
assumptions upon which this rough estimate is based may be criti-
cized. Beach acreage varies from place to place with the motion
of the tides: there may be substantial turnover at a beach
since most people do not stay a full day; some beaches are much
more densely populated than others; and certainly not everyone
swims at public beaches. But, on the other hand, the numbers
used were extremely conservative, and a number of additional

factors were left out. For example, all the other activities
that are contributing elements of demand for recreational facilities at the ocean, such as picnicking, camping, hiking, boating,
fishing, sailing, and sightseeing were completely ignored in
the analysis. So, while unequivocal statements are not warranted,
we can generally conclude that on the basis of a first-order
calculation, we have *at best* enough public recreational shoreline
in New England to fulfill the demands of our *present* population.

Yet, even this is not always true. Near large cities where
demand is the greatest there are critical shortages of shoreline
recreational facilities. Beaches that are over a hundred miles
away do nothing to satisfy these critical metropolitan demands,
although these faraway beaches were averaged into the above calculation. Also, we must be careful here to distinguish between
the *demands* and the *needs* of the population for shoreline recreation, since experience has always shown that new facilities
are quickly used to capacity, indicating an excess of "potential"
demand. While we have no way to measure these "potential" demands, all indicators seem to be that they are quite large!

In the end, what all this says is that, if everyone were
highly mobile and we were clever enough to distribute the supply of shoreline resources efficiently (with ideal transportation
networks), there would probably be enough beach for everyone
today in New England society, given the same patterns of demand.
Sadly enough, neither of these conditions actually exists.
People show definite preferences for nearby, high-quality beaches,
while others who are poorer often have no choice; hence, suitable
areas are filled to capacity, even though one must endure heavy
traffic to reach them. It is clear that we do not have an adequate
supply due to a number of limiting factors such as inadequate
transportation facilities, pollution, and an overall shortage
of beaches.

All this is to say nothing of the future. We have shown
that the demands are growing at a breakneck pace, and that the

supply, limited to begin with, is shrinking steadily. How can
we expect to satisfy the demands of the future when we are having
trouble supplying that which is needed today? And all this with
practically no shoreline left to do anything with? We must
begin to look for a way out of this apparent dilemma, and the
first logical step would seem to be to search out all the remaining sites that might be used for public recreation, and take
immediate steps to preserve them for that purpose. In the remainder of this section we examine two such sites in heavy-demand
areas--Cape Cod and the Boston metropolitan area.

2. Cape Cod

Cape Cod is the favorite vacation spot for Massachusetts
and for much of the Northeast. Every summer weekend sees thousands upon thousands of visitors flowing to and from the Cape,
even though traffic can frequently become almost unbearable.
The economy of the Cape is based heavily on the recreational
business and faces economic problems that are typical of resort
areas. Projections of summer visitors to the Cape for the year
1980 are shown in Table 3.6 and indicate a continuing increase
in the demand for recreational facilities. There is a critical
need to provide for these increased recreational demands, much
of which come from the heavily-populated Boston and Rhode Island
metropolitan areas, while maintaining the character of the area
and the stability of its economic life. The greatest threat is
of overdevelopment and of unplanned sprawl. Additional facilities must be planned and information services provided to help
distribute visitor flows more evenly and efficiently.

The Cape Cod National Seashore is the largest single tourist
attraction on the Cape. Opened in 1963, it consists of 27,000
acres along the northeastern shoreline. Ten percent is to remain
in private ownership, while 30 percent is now federally owned and
the remainder should be acquired quickly (as recommended by the
director of the Bureau of Outdoor Recreation) in light of rapid
land-price escalation. Visitors to the seashore are estimated
to be 20,000 per day by 1980, when it will be one of the largest

Type of Visitor	1960	1980
Summer residents	126,000	237,000
Visitors using motels (average day)	10,400	24,300
Visitors using hotels	4,300	4,300
Visitors using campsites	8,200	19,800
Visitors not using over-night facilities (average day)	37,000	64,000
Visitors not using over-night facilities (peak day)	70,300	106,400

Source: Blair and Stein, Cape Cod 1980 (1963).

Table 3.6 Projections of Summer Visitors to Cape Cod--1980

single employers in the area, providing 50 all-year-round jobs with a $400,000 annual payroll. From now until 1980, 10 million dollars of construction is planned.

On Cape Cod the principal attractions are beaches, as shown in Figure 3.1. The Outdoor Recreation Resources Study for the Massachusetts Department of Natural Resources found beach use on a peak summer day in 1962 to be one-half of the total peak summer day population. Assuming that this proportion will continue, a doubling of beach capacity will be necessary by 1980. The major need is for beaches to serve day trippers who are already overcrowding town beaches. These beaches should be large with adequate parking and easy access. The National Seashore will expand to meet the needs of the lower Cape and the state beach at Scusset should accommodate visitors to the upper Cape. In 1962, Scusset beach attracted only 25 percent of capacity due partly to cold water, but also because of poor publicity. Expanded use of this site could take much of the pressures from some of the more familiar town beaches.

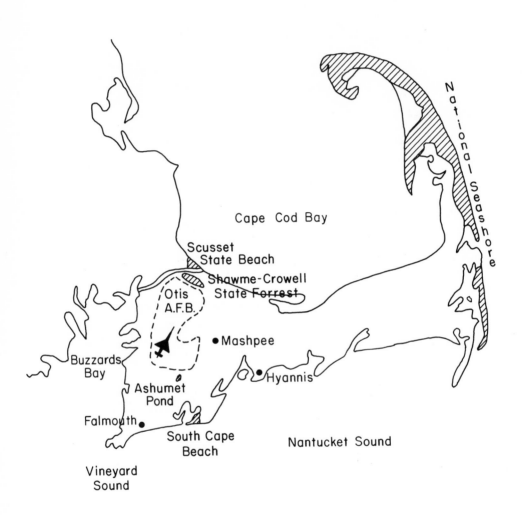

<u>Figure 3.1</u> Cape Cod, Massachusetts

On Buzzards Bay and Nantucket Sound east of Hyannis, no
large public beaches exist although the water temperatures are
much warmer than on the northern side of the Cape. Much more
swimming and related beach activity could be done on the south
side where some beach areas have been taken over for boating
access facilities. On the southern shore, South Cape Beach
in Mashpee, shown on Figure 3.2, has been cited in several studies
as an excellent area for state development. Reports by Blair
Associates[82] and Edwards and Kelcey[83] (prepared for the Department
of Natural Resources) recommend that the state develop a warm
water beach at this location. The DNR studied the area again
in 1968[84] and concluded that South Cape Beach is the last major
opportunity to provide a warm water beach on Cape Cod. This
area is the last piece of Massachusetts coastal land relatively
undisturbed by man. Across Dead Neck, Washburn Island would provide
a fine companion area which could be developed.

The expected growth of Falmouth and Mashpee will create
additional needs even without the increased pressure of day visitors. South Cape Beach is within easy driving distance of the
major Massachusetts population centers and can satisfy at least
one-fourth of the state's needs for swimming and 12 percent of
the beach acreage needs.

South Cape Beach, shown in Figures 3.2 through 3.6, is
essentially a peninsula surrounded by Nantucket Sound and Waquoit
Bay. The beach is flat, occasionally broken by low dunes with
sparse vegetation of dune grasses, bayberry, and wild roses. The
area behind Sage Lot Pond is gently sloping and heavily wooded.
With the exception of access roads and parking, the area is totally undeveloped. South Cape Beach has great recreation potential. The beach, Waquoit Bay, and dune areas could provide a
unique variety of recreation in an undisturbed setting. The
beach proper is composed of 190 acres of beach and low dunes,
10,000 feet along Vineyard Sound and 5,000 feet along Waquoit
Bay. One-half mile of the beach southeast of Sage Lot Pond is
ideally suited for development as a major beach facility with

parking, sanitation, and clothing change facilities. Fifty
acres of beach could be devoted to intensive use there. The remainder of the beach might be left natural to be used under limited supervision.

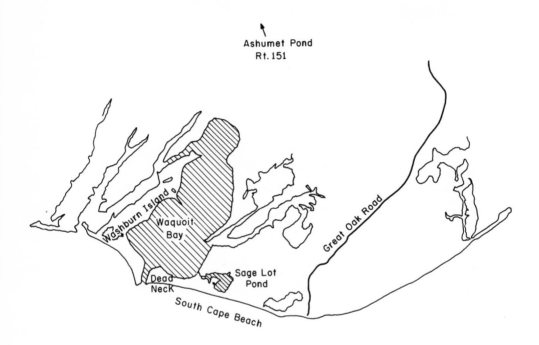

<u>Figure 3.2</u> South Cape Beach and Vicinity

The area bordering the bay provides 3,000 feet of protected
beach. Waquoit Bay is high in shellfish production and good for
other fishing. Migratory waterfowl abound in the spring and
fall. The major recreation activities in this area are fishing,
hunting, nature walks, and boating. The area north of the pond
provides opportunities for picnic facilities and riding. Camping
offers another possible use.

The Department of Natural Resources study team decided that

a buffer area between the state facility and adjacent private land
would be needed. Based on the determination that the South Cape
Beach area would provide a greater variety of outdoor recreation

Figures 3.3, 3.4, 3.5, 3.6 South Cape Beach

opportunities than just swimming, 402 acres would have to be
acquired. Great Oak Road would be improved and access from
Route 151 would pass by Ashumet Pond (where 100 campsites could
be developed), down Washburn Island, cross Dead Neck,
move along the shore of the beach and then inland through Mashpee
to Route 28. Finally, inland ponds can also provide space for the
needed expansion of water-related facilities on the Cape. Blair
and Stein Associates found that there were thirty-five Great Ponds
that were almost completely undeveloped.

3. Metropolitan Boston

Within the coming years the demands for shoreline recreation
in the metropolitan Boston area will far outstrip the supply of
suitable facilities. The drawing power of beaches in this area,
especially in the harbor, is substantial. Almost one out of
every five trips to the metropolitan beaches is made by someone
living more than forty minutes away.[85] This great popularity has

led to gross overcrowding at every beach.

> In 1965 all of the Boston swimming areas combined could
> only accommodate 11,100 bathers. The number of persons
> on an average weekend in the summer desiring access to
> swimming areas will reach 49,000 in 1970.[86]

It appears that the Harbor islands are the last remaining open
areas with access to water that are available for development.
The capacity of Crane's beach to the north could be doubled by
the addition of another parking lot, but this addition is only
a small increment in view of the great needs. Although the
islands are valuable for other uses, they are the only land left
suitable for a variety of recreational activities.

> The Harbor lends itself beautifully to recreational
> and open space development--beaches, boating, fishing,
> clamming, hiking, cycling, and camping....each island
> [has] a personality of its own. Given the rapidly
> growing demand for recreation and open space and the
> existence of alternative sights for housing, airports,
> and industrial development, plans which treat the Harbor
> with respect should fare well under careful economic
> evaluation of the alternatives.[87]

Recognizing this unique suitability, it is crucial that the now
underutilized islands be opened for public recreation and developed as a unit by one agency.

The harbor is central to an area containing 2.5 million
people. Arthur D. Little, Inc., projections for the region
show a population of 3.3 million by 1980 and 4.4 million by 2000.
By 1980, 30,000 bathers per summer weekend day are expected, and
90,000 by the year 2000. Assuming an allocation of 75 square
feet of beach per person, this demand could be handled by an
additional two miles of beach in 1980 and six miles more by 2000.
The Boston Harbor Commission study found the 1990 metropolitan
demand would be for facilities to accommodate 300,000 people
for swimming, 15,000 for boating, 10,000 for camping, and 40,000
for hiking, fishing, and picnicking.

Figure 3.7 shows the Boston Harbor islands and their acreage. A number of plans have been proposed for their development.
Boston Mayor Kevin White has a plan for the development of the

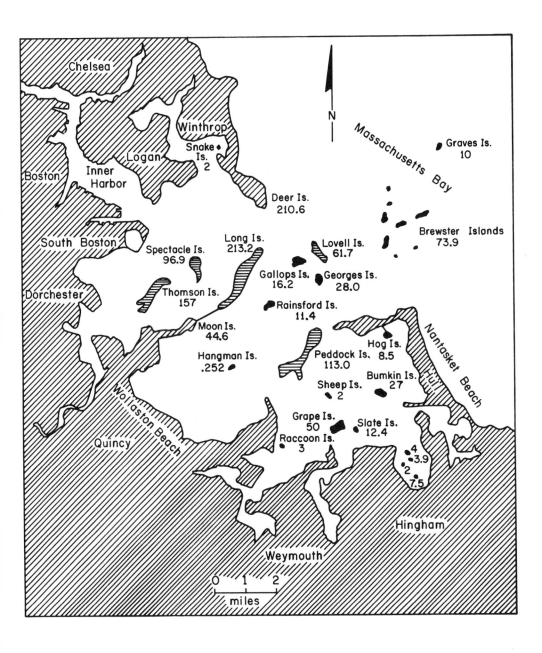

Figure 3.7   The Boston Harbor Islands
(acres)

islands and related shoreline. Its purposes are to transfer privately owned islands to public ownership under the control of one
agency, the Boston Harbor Development and Conservation Corporation within the Department of Natural Resources. This corporation would have the power to acquire and plan a program for recreation and conservation, to develop areas, and to provide ferry
service or bridges for easy access. The outer islands would
be preserved for conservation and light recreation, while the
inner islands (Long, Thompson, Spectacle, and Deer) would be developed for intensive recreation. The second bill would provide
for a Boston Inner Harbor and Industrial Development Corporation
charged with the development of waterfront areas, and industrial
development of specified portions of the city. The corporation
combines the land acquisition powers and the tax-exempt privileges of a public body with the financial capabilities of a private corporation. This body would be able to issue up to $75
million in bonds.

This legislative package is in basic agreement with a bill
proposed by Senator Moakley of the Massachusetts Senate which
would have the Department of Natural Resources acquire and develop the islands for recreation and conservation. Such a bill[88]
has recently been passed by the Massachusetts legislature providing for the acquisition of sixteen islands in Boston Harbor by
the Department of Natural Resources. The bill authorizes an
appropriation of 3.5 million dollars to be expended to acquire
these islands for the purposes of conservation and recreation.
This is an all-important first step in the preservation of these
lands for future use. The islands to be acquired are: Thompson,
Spectacle, Peddocks, Gallops, Bumpkin, Greater Brewster, Middle
Brewster, Outer Brewster, Calf, Little Calf, Green, Raccoon,
Hangman, Grape, Slate, and Sheep. Also, Senator Kennedy has
introduced national legislation for the establishment of a Boston
Harbor National Recreation Area. Currently the bill is in the
U.S. Senate Committee on Interior and Insular Affairs and imple-
mentation must be seen as far in the future. Table 3.7 lists
the harbor islands by present ownership and potential use.

Island	Acreage	Owner	Assessed Value	Potential Use
Long	213.2	Boston, USA	$8,175,800	Intens. Recreation
Deer	210.6	Boston, USA, MDC	4,052,200	Intens. Recreation, Sewage Treatment
Spectacle	96.9	Boston (3/4) Private (1/4)	426,000	Intens. Recreation
Moon	44.6	Boston	291,500	Recreation
Rainsford	11.4	Boston	49,800	Recreation, Conservation
Thompson	157.0	Private	1,069.900	Intens. Recreation Private School
Lovell	61.7	MDC, USA	76,600	Recreation, Conservation
Georges	28.0	MDC	66,400	Recreation
Peddocks	113.0	Private	103,140	Private Homes, Recreation, Conservation
Bumpkin	27.0	Private	10,560	Recreation, Conservation
Greater Brewster	23.1	Private	8,250	Conservation
Outer Brewster	17.5	Private	5,880	Conservation
Calf	17.2	Private	5,140	Conservation
Middle Brewster	12.0	Private	5,080	Conservation
Hog	8.5	USA	44,460	Recreation, Conservation
Green	1.8	Private	800	Conservation
Little Brewster	1.5	USA	–	Conservation
Little Calf	0.8	Private	140	Conservation
Langlee	4.0	Hingham	1,000	Recreation, Conservation
Sailor	3.9	Hingham	1,000	Recreation, Conservation
Ragged	2.0	Hingham	1,000	Recreation, Conservation
Button	0.75	Hingham	1,000	Recreation, Conservation
Raccoon	3.0	Private	1,000	Recreation, Conservation
Hangman	0.25	No Record	–	Conservation
Grape	50.0	Private	2,500	Recreation, Conservation
Slate	12.5	Private	–	Recreation, Conservation
Sheep	2.0	Private	500	Conservation
Snake	2.0	Winthrop	3,700	Recreation, Conservation
The Graves	10.0	USA	–	Conservation
TOTAL	1,152.3		$14,403,350	

Source: U.S. Department of the Interior, Federal Water Quality Control Administration, "Report on Pollution of the Navigable Waters of Boston Harbor" (1968).

Table 3.7 The Boston Harbor Islands

The Metropolitan District Commission is planning for further
development of Lovell and Georges Islands, which it owns. Fort
Warren on Georges Island is open in the summer and 67,000 people
a year visit this area. Peddocks Island, the second largest
in the harbor with 130 acres, could be used for camping, boating,
and general recreation. Currently there are 40 summer cottages,
and owners will be allowed to remain provided private uses are
not inconsistent with the development of the island for public
purposes. Transportation from Boston via ferry service will
be essential. The islands could be linked by ferry service
leaving downtown Boston every half hour with stops at Deer Island,
Long, Lovell, and Peddocks Islands, and Nantasket Beach on the
southern shore.

The Sierra Club is planning a program for the public use
of Lovell Island for the summer of 1970 in cooperation with
the group, Action for Boston Community Development. Lovell,
Georges, and Gallops Islands should be developed further as
a group and linked by pedestrian bridges. The islands are within
a quarter of a mile of each other and the water between them
is less than 30 feet deep. These three have a historical interest
centering around Fort Warren where Confederate soldiers were
imprisoned during the Civil War. Although Georges Island is
open to the public and is very popular, the facilities and the
fort are in poor condition. The Boston Harbor Islands Commission
has recommended the development of a hotel, beach, and protected
boat anchorage in its plan for redevelopment.

Other islands include Thompson, which is privately owned and
used for a boys' academy; Spectacle (which used to be the city
dump); Deer (housing a sewage treatment plant and a prison);
and the Brewsters, which are wild and totally undeveloped. A
focal point for immediate development should be Long Island,
which is owned by the city and is accessible by a causeway. Here
200 acres could provide a variety of recreation without dis-
turbing the hospital located there.

Regulation of the development of the harbor shoreline is

also important. Waterfront districts are needed as part of comprehensive zoning laws. Guidelines for usage and access considerations should be under the control of an agency which could undertake review of construction plans with regional needs in mind.
Mayor White's Boston Inner Harbor and Industrial Development Corporation could serve this purpose, while the current Atlantic
Avenue redevelopment combines residential, commercial, and recreational uses along the waterfront.

Driving for pleasure is the nation's favorite way to spend
leisure time. A scenic road system around the harbor could be
developed, for even today people can enjoy the view while stalled
in traffic on the Southeast Expressway and Morrissey Boulevard.
Visual as well as physical access must be maintained. Selective
acquisition and protection of scenic areas is important. In view
of the present unsightly development of the shore, regulation of
the remaining open areas must be initiated. In most metropolitan
regions, large areas of the limited shoreline are used for industrial, transportation, and residential development with the resul-
tant pollution, noise, and unsightly buildings. Here, where the
demand for water-related recreation is greatest, recreation faces
the most competition for available land. In spite of extensive
development, a surprising proportion of the coastline is open
land. It is run-down, littered, polluted, and often barren, but
it does remain open. Programs must be implemented to assure public use and access.

All but one of the 28 boat-launching areas in metropolitan
Boston are in Weymouth. These are barely sufficient for local
demands now and are woefully inadequate for regional needs. No
public landings exist in the inner harbor. The Boston waterfront
should be freed for development. This area could provide boat
access, parking, and/or various maritime restaurants and stores
to draw tourists. Part of the need can be met by yacht clubs and
private marinas, but there is a critical need for more access and
anchor areas. Development of the harbor islands could provide
more.

Before continuing, it must be noted that an important factor
in the effectiveness of any harbor development plan is the
quality of the water in the harbor. Development of beaches and
other water-related recreation facilities will be a wasted effort
if the water is unhealthy to swim in or even be exposed to, as it
currently is in the areas near most harbor islands. A necessary
companion to any recreational plan for Boston Harbor must be an
effort to reduce the high level of pollution in the harbor. This
topic is treated at length in Chapter 5.

An immediate focal point for development in Boston Harbor
should be Long Island, shown in Figures 3.8 through 3.11. Already owned by the City of Boston, the 200 acres of open, rolling
land are now used only for a hospital for the chronically ill.
The Boston Capital Improvement Program for 1963 to 1975 stated
that the physical plant was in unsatisfactory condition. As
extended care facilities are consolidated on a regionwide basis,
Long Island Hospital will be vacated. Since Long Island is
already accessible to metropolitan Boston by a causeway, development should not wait. Parts of the island could be opened for
public use without hindering the operations of the hospital.

Proposed development of the Dorchester Bay shore of the
island facing the Boston skyline would include areas of intensive and moderate use. Potential swimming beaches could be
incorporated with a boardwalk complex if the water quality were
improved. Fort Hill at the tip of the island would provide
a vantage point for restaurants and a viewing tower. Docks
for a ferry to downtown Boston and for pleasure boats could
be included. The ferry would provide easy access and should
be part of a larger marine transportation service for the harbor
recreation area. The Quincy Bay shore of Long Island with its
meadows and wooded slopes could be developed for 200 family
campsites. Presently, the nearest campground is 20 miles from
Boston in Andover at Harold Parker State Park, already used
to capacity in the summer. Thus, Long Island can offer a unique
diversity of activities in a natural setting. While an immediate

opening is needed, the development must be seen as part of a
comprehensive long-range development of the harbor.   Transfer
of the islands to one agency would facilitate development and

Figures 3.8, 3.9, 3.10, 3.11   Long Island

operation.   Cooperative arrangements with private organizations
could possibly ease some of the burdens of financing.

4. Conclusions

We see that there is much that can be done in the immediate
future to satisfy the increasing demands of the next ten years.
But we must realize that these can only be stop-gap measures,
that in the long-run the problem is much more deeply rooted
in our basic allocative system.   Increased public and governmental
awareness of the uniqueness of the New England shoreline for
recreational use is crucial.   At the very least, the status
quo must be maintained, while any new developments that would
prevent future use of suitable shoreline by the public must
be carefully weighed against the public interest.

REFERENCES

1. U.S. Department of the Interior, Federal Water Pollution
    Control Administration, "The National Estuarine Pollution
    Study," Volume I, Part II, November 3, 1969, p. 27.

2. The George Washington University, "Shoreline Recreation
    Resources of the United States," Study Report No. 4 to
    the Outdoor Recreation Resources Review Commission (ORRRC)
    of the Bureau of Outdoor Recreation, Department of the
    Interior (1962), p. 10.

3. See Reference 1, Volume I, Part I, p. 4.

4. See Reference 1.

5. See Reference 1, Vol. I, Part II, p. 74.

6. Margaret Mead, et al., "Trends in American Living and
    Outdoor Recreation," ORRRC Study Report, No. 22 (1962),
    p. 22.

7. Bayard Webster, "Few Seaside Beaches Left Open to Public in
    Developers' Rush," New York Times, March 29, 1970, p. 54.

8. Ibid.

9. John Bunker, "America's Shoreline is Shrinking," Boston
    Herald Traveler, October 18, 1970, p. 23.

10. See Reference 7.

11. Ian McHarg, Design with Nature, The Natural History Press,
    Garden City, New York (1969).

12. See Reference 2, p. 11.

13. See Reference 1, p. 28.

14. See Reference 7.

15. See Reference 1, Vol. I, Part II, p. 69.

16. See Reference 2, p. 71.

17. Robert C. Cummings, "The Late Great State of Maine,"
    Portland Sunday Telegram, August 30, 1970.

18. See Reference 17; also "Where Went the Maine Coast?"
    August 6, 1970, and "Maine for Sale: Everybody's Buying,"
    August 23, 1970, Portland Sunday Telegram.

19. Pat Sherlock, "The Best of Maine Lost to the Rest of Maine,"
    Boston Globe, September 20, 1970.

20. See Reference 1, p. 1.

21. See Reference 1, Volume II, Part IV, p. 321.

22. See Reference 1, Volume II, Part IV, p. 322.

23. See Reference 2, p. 12.

24. See Reference 2, p. 7.

25. See Reference 19.

26. Herbert Gans, People and Plans, Basic Books, Inc.,
    (1968), p. 109.

27. Ibid.

28. Ibid.

29. Lawrence K. Frank, et al., "Trends in American Living and
    Outdoor Recreation," ORRRC Study Report No. 22, (1962),
    p. 218.

30. See Reference 6.

31. See Reference 26, p. 112.

32. See Reference 29, p. 219.

33. Melvin M. Webber, et al., "Trends in American Living and
    Outdoor Recreation," ORRRC Study Report No. 22 (1962),
    p. 249.

34. See Reference 26.

35. Marion Clawson and Jack L. Kretsch, The Economics of Outdoor
    Recreation, Johns Hopkins Press, Baltimore (1966), p. 31.

36. See Reference 26, p. 112.

37. See Reference 33, p. 245.

38. See Reference 29, p. 220.

39. See Reference 33, p. 249.

40. Department of the Interior, Bureau of Outdoor Recreation,
    "Outdoor Recreation for America," A Report to the President
    and the Congress by the Outdoor Recreation Resources Review
    Commission, January 1962.

41. See Reference 35, p. 5.

42. Marion Clawson, "The Crisis in Outdoor Recreation," American Forests, March/April 1959.

43. Department of the Interior, Bureau of Outdoor Recreation, "Outdoor Recreation Trends," April 1967.

44. See Reference 40, p. 4.

45. Department of Natural Resources, Commonwealth of Massachusetts, "Public Outdoor Recreation" (1954).

46. See Reference 29, p. 224.

47. Harvey S. Perloff and Lowdon Wingo, Jr., et al., "Trends in American Living and Outdoor Recreation," ORRRC Study Report No. 22 (1962), p. 82.

48. See Reference 2.

49. See Reference 2, p. 12.

50   See Reference 2, p. 5.

51. See Reference 40, p. 70.

52. See Reference 2, p. 3.

53. See Reference 2, p. 4.

54. See Reference 1, Vol II, Part IV, p. 116.

55. See Reference 1, Vol. I, Part II, p. 32.

56. See Reference 33, p. 248.

57. See Reference 33, p. 248.

58. Eastern Massachusetts Regional Planning Project, Inventory and Analysis of Recreation, Tourism, and Vacationing in Eastern Massachusetts (1967).

59. U.S. Bureau of Outdoor Recreation, National Recreation Survey, ORRRC Study Report No. 19 (1962).

60. Bowdoin College, Center for Resource Studies, The Maine Coast: Prospects and Perspectives (1966).

61. P. Hendrick, et al., Vacation Travel Business in New Hampshire, New Hampshire Department of Resources and Economic Development (1962).

62. U.S. Department of Commerce, Bureau of the Census, Census of Business (1963).

63. See Reference 58.

64. Blair and Stein Associates, Cape Cod, 1980 (1963).

65. See Reference 2, P. 124.

66. See Reference 29, p. 231.

67. Department of the Interior, Bureau of Outdoor Recreation, "A Report on Land Price Escalation," Jaunary 1967.

68. See Reference 47, p. 84.

69. See Reference 47, p. 86.

70. This discussion is based on the treatment found in J. W. Devanney, III, et al., "Economic Factors in the Development of a Coastal Zone," M.I.T. Sea Grant Project Office Report No. MITSG 71-1, November 20, 1970.

71. See Reference 17.

72. See Reference 17.

73. Gloria Hutchinson, "Harpswell: What Went Wrong?", Maine Times, Vol. 3, No. 2, October 9, 1970.

74. See Reference 17.

75. See Reference 40.

76. "Coastal Zone Management Conference," Hearings before Subcommittee on Oceanography of the Committee on Merchant Marine and Fisheries, House of Representatives, Serial No. 91-14, Volumes I, II, III, October 28, 29, 1969, p. 11.

77. See Reference 1, Volumes I, II, III.

78. See Reference 76.

79. "Coordinating Governmental Coastal Activities," a report by the Task Group on Interagency Coordination, Federal-State Relationships and Legal Problems (COSREL), of the Committee on Multiple Use of the Coastal Zone, National Council on Marine Resources and Engineering Development, September 1968.

80. "Procedures for Evaluation of Water and Related Land Resource Projects," Report to the Water Resources Council by the Special Task Force on Evaluation Procedures, June 1969.

81. See Reference 70.

82. See Reference 64.

83. Edwards and Kelcey, Massachusetts Outdoor Recreation Plan
     (1966).

84. Massachusetts Department of Natural Resources, "Resolve
     Report on the South Cape Beach," 1968.

85. Massachusetts Institute of Technology, Harbor Islands Study
     Group, "The Harbor Islands," prepared for the Boston Harbor
     Islands Commission (1969).

86. U.S. Department of the Interior, Federal Water Pollution
     Control Administration, "Report on Pollution of the Navi-
     gable Waters of Boston Harbor" (1968).

87. The Sierra Club, "Facts about Boston Harbor," No. 4, Decem-
     ber 1969.

88. Commonwealth of Massachusetts General Laws, Chapter 742,
     H.4884.

89. Commonwealth of Massachusetts, Metropolitan Area Planning
     Council, Open Space and Recreation, Vol. II.

CHAPTER 4

CONTROLLING SULFUR OXIDE EMISSIONS

by

Dennis W. Ducsik

Contributing Authors: Larry Donovan
                       Steven Milligan

ABSTRACT

The past decade has marked the emergence of pollution problems as serious matters for public concern. We have come to
realize that our air and water resources by no means have an unlimited capacity to absorb wastes without posing threats to the
health and well-being of American citizens. It is now clear that
our air and water masses, rather than being free goods (available in unlimited quantities of the desired quality), are scarce
indeed and must have their value clearly articulated. Yet the
private market has never been adjusted accordingly; hence, pollution has continued to worsen since appropriate costs have
never been imposed on those who utilize (and degrade) the air
and water.

One important component of the overall air pollution problem
is the emission of sulfur oxides, perhaps the most harmful of all
pollutants. The presence of sulfur oxides in the air has been
found to have adverse effects on visibility, inanimate objects,
plants and animals, and human health. The potential seriousness
of the threats to human health alone is sufficient to merit an
intensive campaign to reduce the levels of these noxious emissions.

We have examined the technology of sulfur oxide abatement,
the supply and demand for sulfurized fuels, and the desirability
of alternative schemes for collective action to control sulfur
oxide emissions. Within this overall context, we have concluded
that the most effective policy would be a staged strategy over
time, utilizing a fuel-tax at the federal level as a short-term
solution and emission fees at the state and local levels in the
long-run. We feel that, if properly formulated, such a policy
can realize the efficiency-seeking advantages of each scheme
while avoiding the shortcomings that preclude sole reliance on
either alternative.

CHAPTER 4

CONTROLLING SULFUR OXIDE EMISSIONS

I. INTRODUCTION

Since the earliest times, men have found the air and water
bodies of their physical environment to be natural receptacles
for the disposal of waste. The capacity of these resources to
absorb and disperse the by-products of civilization in a harmless
way was indeed great, so great in fact that it was seldom (if
ever) thought to be anything other than infinite. There has historically been little or no recognition as to the effects of
changes in the natural environment on the health and well-being
of mankind. Those pollution problems that did occur were viewed
as incidental abnormalities in need of some form of corrective.
action. Not until the twentieth century did man begin to realize
the serious and widespread consequences of his activities, yet
the mounting crisis of the environment remained obscure in the
turbulence of an era of global warfare and economic depression.
After the Great Depression and World War II, emphasis was placed
on the rebuilding of a healthy American society, on growth and
progress toward a high standard of living. It has been this
growth that has brought to the forefront today, for the first
time in history, the real proportions of the environmental crisis.

In recent years, the problems of air and water pollution
have been a focal point of national concern, receiving much
attention in all the media. It is well known that the sheer
number of people and our level of national wealth combine to
generate vast amounts of waste products each year. Our environmental predicament can be traced ultimately to this combination
of soaring *population* and large gains in *productivity*, which together have increased the nation's output enormously in the post-
war period. With productivity growing at about 3 percent per
year and the labor force increasing at a yearly rate of 1 percent,
it is necessary that the economy should grow at 4 percent per
year in order to keep the available capital and labor employed.
The nation's output grew by $100 billion from 1949-1957 and by

$300 billion from 1957-1970. By compounding this 4 percent
growth rate it will be 50 percent larger than it is now by 1980,
an increase of about $500 billion in just ten years. Our pollu-
tion problems are a direct result of this increase in output.
Although we have been pleased with the corresponding increase
in our standard of living, we have failed to recognize that
the damages to society from an activity like pollution are not
reflected in the indices by which we measure our level of national
well-being. As a result, we have been cheating ourselves--at
the expense of the environment--to obtain higher goals of national
output and (what seemed to be) a better standard of living!

Some scientists have suggested that, if our population had
stopped growing about 1850, there would be little or no perturba-
tion of the regenerating capabilities of environmental systems.
Man and nature could have lived in relative harmony. But such
has not been the case: our numbers are doubling every fifty years
or so while growing in wealth and productive capability. As
a result, society now lives with frequently intolerable environmental conditions--air that is not fit to breathe, water that is
not fit to swim in or even sit near, and landscapes that are not
fit to look at and enjoy. We have indeed become the "effluent"
society! Recognition of this has prompted the Council on Environmental Quality to term 1970 "the year of the environment:[1]

> ...1970 marks the beginning of a new emphasis on the
> environment--a turning point, a year when the quality of
> life has become more than a phrase; environment and pollu-
> tion have become everyday words; and ecology has become
> almost a religion to some of the young. Environmental
> problems, standing for many years on the threshold of
> national prominence, are now at the center of nationwide
> concern. Action to improve the environment has been
> launched by government at all levels. And private groups,
> industry, and individuals have joined the attack.

The analyses presented in this chapter and in Chapter 5
represent our willingness to join the fight against the threats
to environmental quality. We have found the problems of air and
water pollution to be critical areas in need of strong public
action. Recognizing that two of the major contributors to these

forms of pollution are 1) industrial and commercial enterprises,
and 2) municipal activities, we have chosen in this study to
focus on two particular environmental problems that can be attributable to these sources: 1) air pollution caused by sulfur compounds emitted into the air by large industries and other users
of fossil fuels, and 2) water pollution in Boston Harbor caused
by the dumping of inadequately treated sewage by the municipali-
ties. Our objective is to present some concrete proposals for
action to alleviate these problems. The remaining sections of
this chapter deal with the problems of air pollution, with particular emphasis on sulfur oxide emissions, while the discussion
of water pollution in Boston Harbor is found in the following
chapter.

II. BACKGROUND ANALYSIS

1. Economic Aspects

We have seen in the discussions in Chapter 1 that certain
goods and services have characteristics which render the classical
functioning of the private market system unworkable or undesirable.
This provides a justification for public concern at least and,
in many cases, for some form of collective action in the public
sector. A good that can be so characterized has been termed,
in a generalized way, a "public good."

The past decade has marked the emergence of pollution problems as serious matters for public concern. We have come to
realize that our air and water resources by no means have an unlimited capacity to absorb wastes without posing threats to the
health and well-being of American citizens. We have found that
the private market has not made the proper adjustments to provide
for the most efficient and beneficial allocation of these scarce
resources. It is now clear that our air and water masses, rather
than being *free* goods (available in unlimited quantities of the
desired quality), are scarce indeed and must therefore have their
value clearly articulated.

Air and water pollution occurs largely because appropriate costs have never been imposed on those who utilize these resources. The desire at every private level to minimize costs has combined with the traditional notion of air and water as free commodities in bringing about a serious misuse of these natural assets. *Yet substantial costs have accrued to others in society who are not involved in the production and consumption activities that make free use of air and water.* In this respect, pollution is often considered a classic example of the breakdown of the private market due to its inability to handle the *side effects* (externalities, external diseconomies, spillovers) associated with the production or consumption of goods and services. As we have shown in Chapters 1 and 3, an important condition for markets to function properly (i.e., bring about an efficient allocation of resources) is that *the total social benefits of consuming a particular good must exceed the total social cost of lost opportunity, which must be reflected in the price of the good.* The existence of uncompensated side effects constitutes a *violation* of this principle. These effects come about when the production (consumption) of certain goods affects other decision-making units which are not doing the producing (consuming). The costs of side effects are *not* included in the price of the good since there is generally no mechanism by which these external costs of society can be returned to the producer as the cost of a factor input to production.

Private industries, municipal governments, and individual persons all contribute to pollution and create the externalities associated with it. Take, for example, the case of a large steel manufacturer who is in need of a new furnace and is faced with a choice between two different models. One alternative provides for a more complete combustion of the fuel used in the steel-making process (thereby substantially reducing the amount of unburned particulate matter discharged into the atmosphere), but is moderately more expensive than the other. Since the market price of pollution is zero, no pollution costs enter into his private

cost-benefit calculations. Hence he will purchase the cheaper
furnace. However, this action may not be without cost to those
residing in nearby communities. Smoke contributes to the rapid
deterioration of house exteriors and leads to larger cleaning
bills; particulate matter irritates the eyes and throats of the
residents; gaseous wastes such as SO_2 and benzyprene may cause
cancer and other disorders in the lungs; and the entire discharge
may contribute to "smog" when meteorological conditions are right,
intensifying the adverse effects mentioned above and possibly
damaging the trees and plant life in the area. Hence, *the air is
not free to society* since residents of the communities involved
must expend resources (or suffer disbenefits) as a result of its
use by another party—the steel mill. Yet there is no market
mechanism to transfer these costs to the steel manufacturers at
the mill!

Consider also the example of a large lake whose waters are
bordered by a number of cities and towns. These municipal governments might all contribute to the polluting of the lake because of
insufficient sewage treatment facilities and antiquated sewer
systems (which combine sanitary with storm drainage that dumps
raw sewage into the lake under the overflow conditions of a heavy
storm). For a single municipality, the installation of new
sewers and advanced treatment plants might not substantially affect the overall pollution level of the lake, yet that town would
incur substantial costs. Consequently, there is no incentive for
individual towns to take any steps to control their contributions
to the lake's pollution. On the other hand, those who normally
swim in the lake or depend on it for their livelihood through
fishing now bear the diseconomies. A firm that uses the water
for industrial purposes may now have to install a treatment plant
of its own (at great cost) to obtain the water quality needed
for its purposes. Again, many people pay the price for water
pollution, but no abatement action is taken since the decision-
making entities that benefit from polluting activities do not
also bear the full costs which result.

A final example is the case of the individual auto owner.

Faced with the option of buying an anti-smog device (at extra cost) for his car, this man considers two situations. If everyone buys the device, the air will be much cleaner; if no one buys, the situation will not improve. In *either* case, such a person would perceive that his little contribution, taken alone, has no significant effect on the overall problem. Why then should he incur the added expense of the device? There is no motivation under any circumstances for the consumer to purchase such a device, and no motivation for a producer to supply it. Yet there are serious external costs involved with the overall air pollution problem to which the sum of all auto owners contribute between 50 and 60 percent. Again, no mechanism exists to transfer the costs to the proper sector, in this case, those who benefit from the use of their autos in a polluted area.

The crucial point that must be reemphasized is that frequently the total opportunity costs to society are *not* reflected in the price of certain goods. *Although the true social costs of having an individual consume/produce a particular commodity may exceed his private benefits, he will base decisions only on the relative weights of his private benefits and private costs.* A good illustration is the case of an individual auto owner who is trying to decide whether to drive from his suburban home to his downtown office. He weighs the cost of driving and the personal inconvenience of traveling on congested, noisy, polluted highways against the door-to-door convenience of this means of transportation. However, his driving to town adds an incremental amount to the congestion, noise, and pollution, all of which has a cost in terms of added inconvenience to the accumulation of *other* motorists and to residents along the route. Yet this cost is not weighed in his individual decision process. If other costs were weighed, the number of motorists would decline until the marginal benefits of driving into town would just equal the marginal costs (to society). In actuality, there are probably too many motorists and too little clean air and quiet surroundings. So we see that the private market, left alone, tends to produce too many *private* goods and too few *public*

goods. This happens because the public goods are *undervalued*
within the private market and are unable to compete on an equal
footing with other goods in the allocation of scarce resources.
For this reason, government might find it desirable to step
in and initiate some form of collective action in order to maintain
social balance and achieve an efficient resource allocation
consistent with the overall goals and values of society!

The preceding considerations have established a useful
framework from which the problems of pollution can be attacked,
since the causes of these problems are rooted in our economic
system. However, the fact that the causes of pollution can be
identified through economic analysis does not necessarily imply
that solutions are to be found in economic policy alone. The
question of environmental quality has very strong technological
and political aspects that must be carefully considered before
effective policies can be formulated. With this in mind, we can
now move on to a more specific discussion of one aspect of air
pollution, the sulfur oxide emission problem. Our aim is to look
at the interacting factors that are relevant to the determination of an appropriate public policy in this regard.

III. SULFUR OXIDE EMISSIONS AND THEIR EFFECTS

1. General

The emission of sulfur oxides into the atmosphere is predominantly a direct result of human activities--their only
natural source is believed to be volcanic gases. Sulfur dioxide
is by far the most common, while other forms of sulfur (such as
sulfur trioxide, sulfuric acid, and sulfate salts) all exist in
the air to a much lesser degree. Data presented by the Department of Health, Education, and Welfare in a 1969 Report entitled
"Air Quality Criteria for Sulfur Oxide"[2] revealed the levels and
sources of these pollutants:

> In 1966, an estimated 28.6 million tons of sulfur dioxide
> were emitted to the atmosphere, as compared with 23.4
> million tons in 1963. The principal share, i.e., 58.2
> percent, came from the combustion of coal, most of which

was used to generate electric power. The combustion of
residual fuel oil and other petroleum products accounted
for 19.6 percent of the total, while the remainder came
from the refining of petroleum (5.5 percent), the smelt-
ing of sulfur-containing ores (12.2 percent), the manufacturing of sulfuric acid (1.9 percent), the burning of
refuse (0.4 percent) and the burning or smoldering of
coal refuse banks (0.4 percent).

Paper-making and some other industrial operations also
contributed minor amounts to the total. In all of these
processes, small amounts of sulfur trioxide or sulfuric
acid are emitted also.

These results indicate that about *80 percent* of the total
yearly sulfur dioxide emissions come about through the combustion
of coal and oil, which contain inorganic sulfides and sulfur-containing organic compounds. In addition, the combustion process
creates one part of sulfur trioxide to approximately 30 parts
sulfur dioxide. In the presence of moisture sulfur trioxide is
converted to sulfuric acid which, along with other sulfates,
constitutes anywhere from 5 percent to 20 percent of the total
suspended particulates in urban air.[3]

Concern over the effects of sulfur oxide pollution has focused on two areas: the global effects of sulfate particles,
which constitute the largest single artificial source of particles; and the more localized adverse effects on the health and
well-being of urban residents. The importance of the global
effects have recently been examined and reported by a recent
conference on the Study of Critical Environmental Problems
(SCEP):[4]

Particles in the troposphere can produce changes in
the earth's reflectivity, cloud reflectivity, and cloud
formation. The magnitude of these effects is unknown,
and in general it is not possible to determine whether
such changes would result in a warming or cooling of
the earth's surface. The area of greatest uncertainty
in connection with the effects of particles on the heat
balance of the atmosphere is over current lack of knowledge of their optical properties in scattering or ab-
sorbing radiation from the sun or the earth.

While data on the global importance of sulfur dioxide emis-

sion is essentially inconclusive, the deleterious effects of
these emissions on the health and well-being of urban residents
are relatively well-established. These effects constitute the
primary cause for concern over the increasing levels of sulfur
oxides in our nation's air.

2. Adverse Effects of Sulfur Oxides

The presence of sulfur oxides in the air has been found to
have adverse effects on visibility, inanimate objects, plants and
animals, and human health, either through its own action or in
combination with other pollutants. These effects have been documented by the previously cited HEW report, "Air Quality Criteria
for Sulfur Oxides."

Visibility

Particles suspended in air can reduce visibility by the absorption and scattering of light from an object and its background.[5] The scattering of light in and out of the line of sight
illuminates the air between an object and the viewer, and the
diminution of visibility is greatest when the particle radius
is in the order of 0.1 micron to 1 micron. Sulfuric acid and
other sulfates constitute from 5 to 20 percent of the total suspended particulate matter in urban air, and about 80 percent
(depending on the humidity) of these by weight are smaller in
radius than one micron. Hence, suspended sulfates make a significant contribution to the diminution of visibility in urban
areas. At a concentration of 0.10 ppm of sulfur dioxide with a
comparable concentration of particulate matter and relative humidity of 50 percent, visibility may be reduced to about five
miles.[6]

Inanimate Objects

It is well known that polluted air containing sulfur oxides
and particulates has adverse effects on a wide range of inanimate
objects, although it is often difficult to separate out the relative contributions of each element. These effects include: increased corrosion rates of iron, steel, and zinc; damage to all

kinds of electrical equipment; damage to building materials such
as roofing, limestone, marble, concrete and cement; deterioration
of textiles such as cotton, rayon, and nylon; and discoloration
or fading of dyed goods. These and other effects are brought
on mainly due to the production of highly-reactive sulfuric
acid, while the extent of the damage is related to the relative
humidity, the temperature, and the presence of other pollutants.
A mean SO_2 concentration of .12 ppm, together with a high particulate concentration, may increase the corrosion rate of steel
by 50 percent.[7]

Plants and Animals

Absorption of sulfur dioxide has been observed to produce
both acute and chronic leaf injury to plants, and it is suspected
that plant growth and vitality might be suppressed even without
any visible damage. Sulfuric acid droplets in polluted fogs may
also damage leaves. The sensitivity of vegetation to damage
from these effects is generally related to temperature, relative
humidity, soil conditions, nutrient supply and other environmental factors. Chronic plant injury and excessive leaf drop may
occur with an annual mean SO_2 concentration of .03 ppm.[8]

Sulfur dioxide and sulfuric acid have been observed to irri-
tate the respiratory system of various animals such as dogs and
cats, causing a detectable increase in airway resistance at certain concentrations. To produce pathological lung change or
mortality, however, relatively high concentrations (compared to
current pollution levels) are required.

Human Health

The effects of air pollutants on human health have been
studied using two approaches: in the *laboratory*, where attempts
are made to establish direct causal links between pollutants and
human health effects; and through *epidemiology*, which looks for a
statistical basis for associating a particular cause with some
effect.

Man responds to sulfur dioxide mainly through *bronchocon-*

striction, or an increase in airway resistance caused by respiratory irritation.

> Laboratory observations of respiratory irritations
> suggest that most individuals will show a response to
> sulfur dioxide at concentrations of 5 ppm ($\sim$14 mg/m^3)
> and above. At concentrations of 1 ppm to 2 ppm an ef-
> fect can be detected only in certain sensitive indi-
> viduals, and, on occasion, exposures to 5 ppm to 10 ppm
> have been shown to cause severe bronchospasm in such
> persons....The exposure of the more sensitive indivi-
> duals to 1 ppm, although it does not produce severe
> bronchospasm, does elicit a detectable response.[9]

Hydrogen sulfide has been found[10] to cause sensory irritation in
individuals exposed to.1 ppm for one hour, while its disagreeable
odor may affect the appetite of sensitive persons at about 5 ppm.
Loss of smell has been reported for exposure to 100 ppm lasting
from 2 to 15 minutes. Sulfuric acid mist with a concentration
of about .03 ppm has produced a respiratory response in humans[11]
when the average particle size is one micron (which is common).
Larger droplets produce sensory irritation (without other physio-
logical effects) at this concentration, but a mist level of about
2 ppm for a few minutes produces coughing and irritation in normal individuals, and might cause acute illness in sensitive
groups over an exposure period of one hour or so.

While laboratory studies have been valuable in generating
information about the relationships between SO_2 and health, their
usefulness in reaching conclusions about ambient air quality criteria is limited by the fact that the experimental environment
does not often simulate actual urban conditions. However,
studies have shown that combinations of sulfur oxides and other
pollutants, such as particulates, may produce effects that are
greater than the sum of their individual effects.

> ...(Laboratory) Exposures have been to high and constant
> concentrations, rather than to the low and fluctuating
> levels commonly found in the atmosphere. Other normally
> occurring stresses, such as fluctuating temperature, have
> not, in general, been applied. These studies do, however,
> provide valuable information on some of the bioenviron-
> mental relationships that may be involved in the effects
> of the sulfur oxides on health. The data they provide on

synergistic effects show very clearly that information derived from single substance exposure should be applied to ambient air situations only with great caution.[12]

A number of epidemiologic studies have investigated the relationship of air pollution to both acute and chronic health effects, especially those occurring in some of the particularly severe air pollution episodes (Meuse Valley, Belgium, 1930; Donora, Pennsylvania, 1948; London, 1952 and 1962; New York, 1953 and 1966). In the London and New York episodes, sulfur oxides and particulate matter have been correlated significantly[13] with the observed effects of increased mortality and morbidity. Other studies[14] (Rotterdam, The Netherlands; Eston, Great Britain; Buffalo, New York; Genoa, Italy; Berlin, New Hampshire; Nashville, Tennessee; Port Kambla, Australia; Chicago, Illinois) have demonstrated that smaller, steady concentrations of SO_2 along with other pollutants in urban air are also correlated with increased mortality and morbidity.

> Analyses of numerous epidemiologic studies clearly indicate an association between air pollution, as measured by sulfur dioxide accompanied by particulate matter, and health effects of varying severity. This association is most firm for the short-term air pollution episodes.
>
> The epidemiologic studies concerned with mortality also show increased morbidity. Again, increases in morbidity as measured, for example, by increases in hospital admissions or emergency clinic visits, are most easily detected in major urban areas.
>
> It is believed that, for large urban communities which are routinely exposed to relatively high levels of pollution, sound statistical analysis can detect with confidence the small changes in daily mortality which are associated with pollution concentrations.
>
> The association between long-term community exposures to air pollution and respiratory disease incidence and prevalence rates is conservatively believed to be intermediate in its reliability. Because of the reinforcing nature of the studies conducted to date, the conclusions to be drawn from the type of study can be characterized as probable.[15]

Table 4.1 lists the conclusions of the Department of Health,

Possible Effects	SO$_2$ Concentration (ppm)	Other Factors Present
Increased Mortality	.52 (24-hr average)	suspended parti-cles at soiling index of 6 cohs or greater
Increased Daily Death Rate	.25 (24-hr mean)	smoke in concen-tration of .26 ppm or greater
Increased Mortality	.19 (24-hr mean)	low particulate levels
Increased Absenteeism Increased Hospital Admission (older persons with respiratory disease)	.11-.19 (24-hr mean)	low particulate levels
Sharp Rise in Illness Rates (patients over 54 with severe bronchitis)	.25 (24-hr mean)	particulate matter
Accentuation of Chronic Disease Symptoms	.21 (24-hr mean)	.11 ppm smoke concentration
Increased Frequency of Respiratory Symptoms and Lung Disease	.037-.092 (annual mean)	.068 ppm smoke concentration
Increased Frequency and Severity of Respiratory Diseases (school children)	.046 (annual mean)	.036 ppm smoke concentration
Increase of Mortality--Bronchitis and Lung Cancer	.040 (annual mean)	.059 ppm smoke concentration

Source: Air Quality Criteria for Sulfur Oxides (Ref. 2, pp. 10:20-21)--in descending order of reliability

Table 4.1 Health Effects of SO$_2$ Concentrations
in Polluted Air

Education, and Welfare regarding the effects on human health of SO_2 emission combined with other air pollutants. From this data, the department concluded:

> ...it is reasonable and prudent to conclude that sulfur oxides of 300 $\mu g/m^3$ (.1 ppm) or more in the atmosphere over a period of 24 hours may produce adverse health effects in particular segments of the population...[16]

In addition, adverse health effects were noted for an annual mean concentration as low as 115 $\mu g/m^3$ (.04 ppm) of SO_2.

3. Conclusions

The potential seriousness of the threats to human health and well-being posed by sulfur oxide emissions in particular and by pollution in general is sufficient to merit an intensive campaign to reduce the levels of these noxious emissions. Other considerations lend support to this contention. For example, a polluted environment may have a depressing effect on urban residents, caused by the perpetual presence of hazy visual conditions or offensive odors. In addition, little is known about the effects of pollutants on the complex interaction of delicate ecosystems such as the biochemical cycles of oxygen, sulfur, carbon, and nitrogen. Yet, it is possible to conceive of ecological cycles "in which the specific toxicity of a pollutant for a single species could cause an entire food chain to collapse, but the extent to which this might happen is unknown. Too little is known of the effects of pollutants on too few species to suggest even how such problems might be attacked."[17] All of these considerations, taken together, represent a strong rationale for a concentrated effort to be launched, on both a local and national scale, to reduce the existing levels of SO_2 emission. There seems to be clear evidence that these emissions constitute a primary causal factor (when combined with other elements) in the adverse effects of air pollution on American society.

Before we can decide which tools of public policy can best be used in attacking the SO_2 problem, it is important to examine the *technology* of sulfur oxide control. This aspect is closely

related to economic considerations and, together with the economics, will determine the applicability and effectiveness of
alternative schemes.

IV. CONTROL OF SULFUR OXIDE EMISSIONS

We have seen that each year nearly 80 percent of the sulfur
oxides emitted into the air result from the combustion of coal and
oil. In 1966, 42 percent of the total U.S. emissions of SO_2 came
from power plants, approximately 23 percent from large industrial
processes, and much of the remaining 35 percent from space-heating
sources, while the trend has been toward large point sources[18]
(except for heating). It is estimated[19] that, in 1971, "over
60 percent of the 44 million tons of sulfur dioxide discharged
into the atmosphere in the U.S. will come from coal- and oil-fired
plants. By the year 2000, when total emission will have increased
to nearly 120 million tons, over 80 percent will result from power
generation." This points to the pressing need to find ways to
bring the harmful emissions of these sources to within acceptable
levels, and all indications are that we must look toward advanced
technologies to provide an answer to the problem. The purpose of
this section is to explore the two major areas of control technology that have been developed to data: 1) the removal of sulfur
from stack gases, and 2) the desulfurization of fossil fuel
(coal, oil). We shall also examine other control techniques,
such as regulation of the use of high-sulfur fuel, the discharge
of fuel gases at high velocity and temperature from stacks, and
the possibility of improvement in combustion efficiency.

1. SO_2 Removal from Stack Gases

While there are presently no stack gas removal processes
currently in widespread use, a number of methods are being studied
carefully. The major factor prohibiting large-scale installation
of these processes is their extremely high costs.

Limestone Injection

This process, which appears to be the closest to practical

use, consists of the injection of limestone, dolomite, or some
other reactive metal oxide into the fuel-burning furnace. These
substances react with sulfur oxides to produce metallic sulfates
that can be removed by dust-collecting equipment. When a dry
collection system is used, less than 50 percent of the sulfur is
removed, while a wet scrubbing system will result in nearly 90
percent removal. The costs for these processes for an 800 megawatt, coal-fired plant at 90 percent load factor are estimated[20]
to be 3.3 million dollars (.29 mills/Kwh) for the dry process and
4.65 million dollars (.35 mills/Kwh) for the wet process. Wet
washing has several disadvantages other than being costly. It
could require large quantities of water which would be discharged
containing effluents; and a wet, nonbuoyant (cold) plume could be
trapped for long periods by severe inversions.

Alkalized Alumina Process

This process, which has been developed by the U.S. Bureau of
Mines, consists of the absorption of sulfur oxides by a metal
oxide, followed by regeneration (with a reducing gas) which produces marketable sulfur. The process is estimated to yield removal efficiencies of 90 percent or more, with the capital cost
estimated[21] to be about 8.6 million dollars for an 800 megawatt
plant. The development of this process has incurred setbacks
recently[22] due to difficulties with the stability of the solid
reacting agent.

Catalytic Oxidation Process

This process converts sulfur oxides in the stack gas to weak
(75 to 80 percent) sulfuric acid, and is well developed based on
the technology of sulfuric acid manufacturing. However, for an
800 megawatt power plant the system is estimated[23] to cost from
16 to 24 million dollars, while the economics would further depend on the sulfuric acid market in any given region.

We should note at this point that, while none of the above
processes are commercially proven and their costs are highly
speculative, there has been sufficient progress to indicate that

reliable technologies should be available in the not-too-distant
future. The additional power costs to U.S. consumers would then
be on the order of .5 to 1 mill/Kwhr. *It should be noted that
this would only solve part of the SO_2 problem, since the technologies are feasible only for large-scale installations such as
power plants and do nothing to eliminate the contributions from
medium- and small-scale sources (such as domestic heating, which
is a substantial contributor to emission in urban areas).* However, as the trend continues toward large point sources, the
availability of an effective technology to remove sulfur oxide
from stack gases seems to represent a very effective solution to
this aspect of the problem. Hence, it is important to insure
that the development of this technology continues at a rapid
pace. But we must be careful not to rely solely on future technological developments, since the seriousness of the SO_2 problem
is increasing at an accelerated rate.

> Several of these processes will doubtless turn out to
> be technical successes, but the economics are not yet
> well established for even the most advanced. Contrary to
> a widely held belief, the technology does not in fact
> now exist to effectively control SO_2 emissions, and it
> is coming along too late to prevent a very substantial
> increase in SO_2 pollution during the next ten to fifteen
> years.[24]

2. Desulfurization of Fossil Fuels

The removal or reduction of sulfur in fossil fuels before
combustion offers another possibility for the effective control
of sulfur oxide emissions. In 1966, the relative contributions
of coal and oil combustion to the total SO_2 emissions are shown
in Table 4.2.

Coal

The combustion of coal for the generation of electric power
and other purposes is the largest single contributor to the SO_2
pollution problem in the United States. Unfortunately, the removal of sulfur from coal is often a very complex and prohibitively
expensive affair. The sulfur is present in two main forms:
organically, in chemical combination with the coal; and mixed in

Source	SO_2 Emissions, Tons	Percent of Total
Utility Coal	11,925,000	41.6
Utility Oil	1,218,000	4.3
Other Coal	4,700,000	16.6
Other Oil	4,386,000	15.3
		77.8

Source: U.S. D.H.E.W., Public Health Service, Control Techniques for Sulfur Oxide Air Pollutants, January 1969

Table 4.2 SO_2 Emissions from Fuel Combustion in 1966

as pyrite, a mineral impurity. Techniques exist for removing some of the pyrite sulfur in coal using a crushing and washing process by which the iron pyrite is separated from the coal through flotation. However, the sulfur in chemical combination is very difficult to remove without breaking up the coal. Some complicated schemes such as hydrogenation and liquefaction are under intensive research, but at present these techniques are extremely expensive. So the extent to which desulfurization of coal can be an effective means of reduction of SO_2 emissions depends strongly on the relative amounts of the two different forms of sulfur present in the coal. This presents another problem--the two forms exist in greatly varying proportions such that it is very difficult to tell which coals may be readily cleaned and which may not. The best estimate is that approximately 25-30 percent of the sulfur content can be removed on the average. It has been suggested[25] that only about 15 to 20 percent of the high-sulfur utility coal is washable to 1.0 percent sulfur, at an additional cost of from 25 to 75 cents per ton.

Oil

In the process of refining crude oil through distillation, the crude is separated into the various petroleum products ranging

from No. 1 distillate (the lightest) to No. 6 residual oil (the
heaviest) as shown in Table 4.3. The sulfur tends to concentrate
more in the heavier parts than in the lighter, where desulfurization is well established and actually part of the normal refining
process. Because of its lower cost, the residual fuel oil (grades
5 and 6) is generally used by large consumers, while greater than
80 percent of this oil contains at least 2 percent sulfur.[26]

Grade	Use	Maximum Sulfur Content (%)
No. 1 distillate oil	Pot-type burners	0.05
No. 2 distillate	General purpose domestic heating	1.0
No. 4	Burners without preheating facilities	No limit
No. 5 residual oil	Burners with pre-heating facilities	No limit (generally 1-3)
No. 6 (Bunker C) residual	Burners with pre-heaters permitting high viscosity fuel	No limit (generally 1-3)

Source: A. C. Stern, Air Pollution, Vol. III, p. 21
         The Academic Press (1968)

Table 4.3 Sulfur Content of Fuel Oil Grades

The reduction of sulfur content in these fuels is more difficult
but is under intensive development and has met with reasonable
success. Desulfurization units in Venezuela are either presently
operated or being planned by Shell Oil, Standard Oil of New
Jersey, and Humble Oil. These units are quite costly, and it is
estimated[27] that the eventual reduction of the sulfur content in
residual oil from 2.6 percent to .5 percent will result in a 20
to 35 percent increase in fuel costs.

In general, the technology for the desulfurization of oil is
much better developed than that for coal, and provides an encour-

aging prospect for the effective abatement of SO_2 emissions from
oil-burning sources, particularly as development continues and
the processes become more economical. However, it appears at
this point that the effective abatement of SO_2 emissions from
coal-burning sources must look to control techniques other than
desulfurization, since processes of this sort are not by any means
ready for widespread application.

3.   Use of Low-Sulfur Fuels

A recent legislative trend at the state and local levels has
been to require the use of low-sulfur fuels in all combustion
processes. While this brings about an immediate reduction in
SO_2 emissions, questions of availability and cost may pose severe
obstacles to the use of such controls. Low-sulfur fuels are produced naturally in mining and refining operations, or can be
generated by the desulfurization techniques discussed previously.
In this section we restrict our discussion to the supply aspects
of low-sulfur fuel that is generated *without* desulfurization.

Coal

The primary remaining coal reserves of the United States
consist of bituminous coal (high rank), subbituminous coal (low
rank), and lignite (low rank). Problems in the burning of low-
rank fuels have led to their limited use in the past, although
recent technological advances[28] have generated increased interest
in these fuels. Table 4.4 shows the distributions of coal reserves, approximately half of which are considered recoverable,
in the United States in 1965. It is easy to see from this table
that, while the United States has an abundant overall resource
of coal, the size and distribution of low-sulfur, high-rank sup-
lies make long-term reliance on this type of fuel impractical for
most parts of the country. For example, the large east coast
market draws on coal reserves east of the Mississippi River
of which only about 16 percent is of suitable rank and low in
sulfur. West of the Mississippi River there is a greater reserve
of low-sulfur coal, but nearly 85 percent of this is of low
rank. Hence, only about 13 percent of the total western reserves

Reserves	East of Mississippi R. (10^6 short tons)*	West of Mississippi R. (10^6 short tons)*
High rank, low (<1% sulfur -bituminous	90,000	140,000
Low rank, low (<1% sulfur -subbituminous -lignite	negligible negligible	387,200 406,000
High (>1%) sulfur -mostly bituminous	431,400	125,400
Total Reserves	521,400	1,058,600

*Figures are rounded and refer to coal in seams at least 14 inches thick and less than 3000 feet deep in explored areas.

Source: U.S. D.H.E.W., Control Techniques for Sulfur Oxide Air Pollutants (see Reference 20)

Table 4.4 Estimated U.S. Coal Reserves, 1965

are of sufficiently high rank and low in sulfur. On a national basis, then, only about 15 percent of the total coal reserves would be suitable from a pollution abatement standpoint.

The relative scarcity of low-sulfur coal is compounded by the fact that nearly one-fourth of it is exported each year, while much of the rest is sold at a premium to the metallurgical coke industry.[29] Unless these patterns change, there can be no significant long-term reliance on the natural sources of low-sulfur coal. This is particularly true for the eastern portion of the country since, even if techniques were perfected for burning low-rank fuels, the supply of low-sulfur coal is limited in comparison with high-sulfur reserves. For the *short run*, however, the reserves of high-rank, low-sulfur coal are quite substantial, on an absolute scale, in both portions of the country. If patterns of consumption were to be altered, these re-

sources would seem to be more than adequate to meet the early requirements of pollution abatement programs for low-sulfur coal.

Oil

As previously noted, the refining of crude oil produces a
number of grades of fuel oil (See Table 4.3) as well as other
petroleum products such as gasoline. Fuel oils are generally
classified into two categories--distillates and residuals. The
distillates (grades 1 and 2) are primarily used for heating homes
and small apartment buildings, domestic hot water, and industrial
processes where simple burning equipment is used. The average
sulfur content of this fuel is between .04 and .35 percent by
weight,[30] thereby contributing an inconsequential proportion to
total SO_2 emissions. These grades of fuel oil, however, are not
practical financially for use by the larger consumers who generate the bulk of SO_2 emissions.

Residual fuel oils (grades 4, 5, 6) are used primarily for
heating industrial and commercial buildings and apartment houses
(4 and 5) and the firing of the large boilers used by utility
companies (6). Table 4.5 shows the total consumption of residual
oil in the United States in 1966 as well as its distribution by
source and sulfur content. Table 4.6 shows the relative usage of
residual oil for heating, power generation, and industrial
operations.

It is clear from Table 4.5 that the principal source of
residual fuel oil, most of which is very high in sulfur content,
is from foreign refineries, particularly in South America.
Nearly 90 percent of all the residual oil consumed in the U.S. in
1966 had a sulfur content greater than 1 percent by weight, while
about 75 percent had a sulfur content greater than 2 percent. As
a result, combustion of residual oils for power generation,
heating, and industrial purposes has played a major role in the
SO_2 pollution problem.

Since residual oil is part of the output of the refining
process, the production of low-sulfur residual implies modifica-

	Total Amount (10^3 bbl)	>1% Sulfur (10^3 bbl)	>2% Sulfur (10^3 bbl)
Imported Residual	376,800	376,800	368,940
Residual from Domestic Crude (high sulfur - >1%)	136,630	136,630	65,740
Residual from Domestic Crude (low sulfur - <1%)	40,740	-	-
Residual from Foreign Crude	59,830	38,900*	26,900*
Total Residual Oil Consumed	614,000	552,330	461,580

*Estimated from additional data presented in Reference 15.

Source: U.S. D.H.E.W., Control Techniques for Sulfur
         Oxide Air Pollutants (see Reference 20)

Table 4.5 Total U.S. Residual Fuel Oil Consumption by
           Source and Sulfur Content (billion barrels (bbl))

Use	Amount (10^3 bbl)	Percent of Total
Heating (apartments and commercial)	167,470	27
Industry (including oil company use)	176,230	29
Power generation	140,600	22
Other (military, railroad, marine)	129,700	22
Total	614,000	100

Source: Same as Table 4.5

Table 4.6 Consumption of Residual Oil by Type of User

tions or additions to that process. Hence, switching from high-
sulfur fuel oil to low-sulfur fuel oil constitutes trying to buy
desulfurized residual from the refining companies. This then is
strictly a question of economics since the technology is fairly
well developed.

4. Other Methods of Control

Other methods that would help control SO_2 emissions are the
use of nuclear or hydroelectric power, combustion of natural gas,
the development of new energy-related technologies, and the use
of tall stacks for better atmospheric dispersion.

Nuclear Power

The widespread use of nuclear power, while having an extremely beneficial effect in reducing the air pollution problem,
has many difficulties of its own, including thermal pollution and
disposal of high-level radioactive wastes. The fact that nuclear
power plants do not significantly pollute the air is by no means
a sufficient condition for primary dependence on nuclear power.

Hydroelectric Power

By the year 2000, there probably wouldn't be enough river
water in the entire country to satisfy the *cooling* requirements
of all power plants, let alone drive their generators!

Combustion of Natural Gas

While new reserves of natural gas are being found, "the
domestic supply of this fuel at current prices will probably become limited before the turn of the century because of increased
production costs."[31] Widespread combustion of natural gas by
large consumers such as power companies is not practical since
the supply cannot be guaranteed on an uninterruptible basis.
Hence, the primary users of natural gas are the smaller residential and commercial users to whom the supply can generally be
guaranteed.

Tall Stacks

The discharge of flue gases at high velocity and temperature from tall stacks has been suggested as a means to reduce SO_2 ground-level concentrations. This appears to have limited applicability, as there are objections on several grounds:

1) usefulness is limited by local meteorological and topographic conditions--there is some evidence of early morning fumigations[32]

2) the stacks would be very costly--over 2 million dollars for a 900 ft stack--and could only be used by large sources of SO_2 emissions

3) tall stacks would create a potential hazard to aviation, especially near airports

4) the proliferation of tall stacks in a very industrialized urban area would be a substantial eyesore.

New Combustion Technologies

The most significant area in which new technology offers the promise for any substantial alleviation of the sulfur oxide problem is in the combustion of coal. By using fluidized-bed combustion in the presence of lime (a desulfurizing agent), boiler efficiency could be improved while at the same time eliminating sulfur from the stack gases.[33] While these developments, if actively pursued, offer potential solutions for the future, there seems little likelihood that they can be of real benefit in the short run.

5. Conclusions

Ultimately, the effective control of sulfur oxide emissions from coal- and oil-burning sources depends on technology. Whether through the removal of sulfur from stack gases, desulfurization of the fuel itself, or new combustion techniques, the ultimate solution to this phase of the air pollution problem is rooted in technical development. This assumes that we are to continue to rely on fossil-fuel combustion as the primary means of electric power generation. Reliance on nuclear power must be regarded as a long-term solution since only a small percentage of the total power generated each year comes from this source. Even then, it is not altogether clear that this would solve more problems than it would create (see Chapter 2).

We have seen that there has been a great deal of research
and development in a broad spectrum of technological areas relating to the SO_2 pollution problem. To find a suitable alternative in the shortest time requires that an incentive exist for
research in promising areas to push ahead with the development
or refinement of appropriate technologies. The private market
does a good job at finding efficient solutions, given that competitors in a free marketplace have the flexibility of decentralized decision-making. This has important implications regarding
the formulation of public policy to regulate the level of sulfur
oxide emission. *The most desirable policy is one that preserves
the freedom of all interested parties to consider a wide range of
technological alternatives*. This provides the needed incentive
for the scientific and engineering establishment to proceed with
the investigation of technologies that may be economically attractive. This is a most important conclusion. While the requisite
techniques are not yet ready for widespread use, many seem to be
right on the threshold. It would be a mistake of severe proportions to discourage the final stages of development in some of
these areas through unenlightened public policy. We must be
careful to consider this point in the establishment of guidelines for the effective control of sulfur oxides, both in the New
England region and throughout the nation as a whole.

V.   ALTERNATIVE SCHEMES FOR COLLECTIVE ACTION

The analysis in Section II described the problems of pollution as basically economic in nature. The most general solution
to these problems can also be described in economic terms, i.e.,
if one desires to maximize society's well-being in *real* terms,
the social costs of pollution should be transferred to producers
as a factor input to production. While this may by itself seem
simple enough, the determination of the proper mechanism by which
this can be accomplished is an extremely complex issue that must
be examined within the relevant political, technical, and econo-

mic context.

In doing so, we should keep in mind two general principles
that are fundamental to the concept of seeking *efficiency* in our
pollution abatement activities. These are:

1)   There is great advantage in *decentralizing decisions*
     regarding the choice of abatement techniques

2)   It is important that *flexibility* be maintained in
     the formulation of public policy

Decentralization

The great advantage of decentralized decision-making is that
individual producers maintain the freedom to expend their least-
valued set of resources in complying with a pollution abatement
program. This provides an incentive to explore a wide range of
potentially attractive technical alternatives that will control
their emissions. Hence, in choosing a policy that seeks efficiency, care must be taken to avoid discriminating against some
technologies that may ultimately prove useful and valuable in the
long-run.

Flexibility

Since the state of the art of pollution control is still
evolving, any abatement scheme should have a flexibility that will
allow it to be revised as (1) new technical capabilities and (2)
more complete informational resources become available. These
considerations, especially the latter one, have too frequently
been overlooked in our initial responses to pollution problems
at local levels. The seriousness of the sulfur oxide problem,
for example, will vary a great deal from one city to the next as
well as with changing meteorological conditions within each city.
While certain abatement measures are justified as emergency steps
to counteract dangerously worsening situations, such measures
are generally lacking in the type of flexibility that is needed
to attain efficiency in the long-run. Yet too often a legislative precedent is set that is observed to have desirable short-
term effects but which may actually be quite undesirable in its

long-term behavior. Hence, we must be careful in formulating
pollution policies to avoid the foreclosure of future options and
the loss of valuable flexibility through the establishment of
a hard-to-change legislative "momentum."

In addition to the concepts of decentralization and flexibility, a new set of variables must now be considered, i.e.,
the *political* factors that affect the implementation of pollution
abatement schemes within the institutional environment. In
this section, we review the major alternative mechanisms that
have been suggested to bring about effective air pollution control,
taking particular care to examine how the applicability of each
alternative is a function of the contextual factors in relation
to sulfur oxide emissions. The basic alternatives are:

1) Direct regulation
2) Economic incentives

1. Direct Regulation

The most common legislative approach to the problem of pollution has traditionally been to regulate it directly through the
legal institutional framework. Direct regulation involves the
use of laws, licenses, permits, registrations, and directives--
based on some compulsory standard--to discourage pollution beyond
a certain level. The appropriate governmental unit would attempt
to determine the "right" or "acceptable" level of pollution emissions and then enforce these standards through some systems of
inspection, legal action, fines, or other means. Recently the
trend has been toward the regulation of fuels and/or equipment.
This approach has been described by Lawrence W. Pollack:[34]

> The establishment of emission standards is considered
> by many to represent the ideal legislative approach, as
> it theoretically leaves to the owner's discretion the
> precise type of equipment or fuel to be used....
>
> Many legislative and administrative bodies, however,
> have long recognized that the bare setting of emission
> standards were not sufficient, and that fuels and equip-
> ment should be directly regulated....New York City recog-
> nized and followed this approach. Among the reasons

cited were that there were too many smokestacks to permit constant observation for visible smoke violations,
and that no practical scientific equipment was available which was capable of being placed and maintained
in every smokestack to constantly record the amounts
of invisible gases or particles being emitted. In any
event, a strict emission standard has the indirect result of requiring a change in either equipment or fuel
in order to meet the standard, for the emission must
depend upon what substance goes in and what is done to it.

While the concept of setting an emission standard is a healthy
one in that it allows the individual decision-maker to decide on
the type of equipment or fuel to be used, the net effect of direct regulation may defeat this purpose since it is subject to a
number of drawbacks in varying degrees of severity.

First, it may be difficult to determine the "threshold"
amount of pollution since little is known about the damages done
by some forms of pollution and the costs and benefits associated
with it are so ill-defined. Not much conclusive information is
available concerning the long-term effects of air pollution on
man's lungs, the ecology of lakes, streams, and other natural
systems, or the global effects of man-made emissions. Clearly
any pollution abatement system must be flexible enough to adapt
to the changing body of knowledge as more extensive information
becomes available, and flexible enough to handle different kinds
of situations. Frequently the amount of "safe" pollutants in
the air depends primarily on prevailing winds and other meteorological conditions in the area. These considerations might be
very difficult to incorporate in direct regulations. On the
other hand, when the pollution levels of urban air are well
above those criteria agreed on by most authorities as being
detrimental to the health and well-being of the general public,
the question of finding an optimal solution must take second
priority to that of taking immediate steps to counteract an
emergency situation. A. V. Kneese has argued that "making improved decisions based on economic data does not necessarily
require that we know the total costs and gains at all."[35] Hence,
this drawback can be considered relatively inconsequential under

many present circumstances, but is certainly of great importance
in the long-run.

A second and much more serious problem with direct regulation is that it can stifle the inventiveness and technological
ingenuity associated with decentralized decision-making in a
profit-incentive system by attempting to regulate the wrong thing.
For example, prohibiting the use of high-sulfur fuel by power
companies does not create any incentive for those companies to
utilize some newly-effective ways of burning the high-sulfur fuel
without polluting, thereby discouraging the development of alternate, lesser-cost control schemes. If the use of high-sulfur
fuel is outlawed unconditionally, then there is no incentive for
researchers to seek better stack gas removal techniques or improve on the combustion process. If one type of burner is required to be used, then there might be no incentive for oil companies to further develop their capabilities to produce low-sulfur
oil.

To avoid this problem, some cities have allowed for *variances* or *exemptions* to be granted if alternative abatement techniques become available.

> The fact that many experiments are now being conducted
> for methods of removing sulfur dioxide from the stack
> led to a novel provision in the New York City law. It
> permits an exemption from the sulfur limitations for an
> operator whose equipment has control apparatus capable
> of continuously preventing the emission of sulfur di-
> oxide greater than would be the result of the direct
> sulfur content limitations....This alternative was es-
> tablished even though no existing method was considered
> economically feasible for commercial operation in this
> country....This exemption provision was obviously de-
> signed to stimulate industry into channeling research
> and development efforts toward new methods of air pol-
> lution control. There would seem to be no legal objec-
> tion to this type of legislation since it is in the
> form of a permissive exemption, and the standards re-
> quired are specifically described.[36]

While the conceptual basis for such a provision is sound, its
value may be negated when we consider a third potential problem
area with direct regulation--that of *efficient administration and*

enforcement. Given that standards and regulations have been set,
regulating agencies must have a workable detection and measurement scheme to uncover violators. Even then, unless penalities
are sufficiently high and quickly applied (not often the case
through the courts), many polluters would rather risk the fine if
they determine that it is cheaper than initiating abatement controls. This "license to pollute" may ease the conscience of the
polluters but does nothing to abate the pollution problem. Another administrative problem involves determination that devices
and controls *remain* effective once in operation. Most automobile
anti-smog devices do not work well at all unless the motor is
finely tuned, a situation which might occur only once or twice a
year with many auto owners. The administrative costs involved in
periodic checking of such devices could be enormous! Yet, if
there is no enforcement, the polluter always has an incentive to
expend as few resources as possible in maintaining the efficient
operation of an antipollution device. This argument applies to
power companies and auto owners alike and is especially pertinent in circumstances where the antipollution operation inter-
feres with other profitable activities. Government policies must
be careful to avoid such situations that tend to nullify the
benefit to be gained in a pollution-abatement program of this
sort.

Finally, the very nature of some operations handled by regulating agencies in the public sector can be an obstacle to effective action. Too often these agencies are controlled by the lob-
bies of the industries they are supposed to regulate. One can
point to the surprising frequency with which people who hold high
offices in certain federal agencies relinquish these posts to
take positions with the industry under control. This problem is
equally common at the state and local levels. Another drawback
is that, even when strong measures are provided for regulation,
bureaucratic inefficiency can sometimes bog down the whole operation and render it ineffective. The worst situation we could
find ourselves in would be to be spending millions of dollars and
not solving the problems we are trying to attack!

Even if all of these difficulties turned out to be resolvable and direct regulation policies attractive in the short-run
(as may be the case at some local levels), some other more
broadly-based difficulties might be encountered in the long-run.
For example, consider the situation if most large cities enacted
laws which called for the use of low-sulfur fuels unless an exemp-
tion is granted. In this case, fuel producers might be reluctant
to proceed directly with the expansion of production facilities
for desulfurized fuel, depending on what the economic picture
looked like for stack gas removal technology in the future. If
they geared up to meet an enormous demand for low-sulfur fuel,
and then a technological advance in stack gas removal gave all
users a variance on the use of high-sulfur fuel, the demand would
switch to high-sulfur fuel, leaving the producer with idle, expensive production facilities for the low-sulfur type. Consequently, if the fuel producers did not make enough low-sulfur
fuel available, then the users would be stuck with the high cost
of violating the emission standards and would still be burning
the polluting fuels.

Although legal regulation (such as limiting the sulfur content of fuel) may seem to be the most immediate and relatively
uncomplicated means to effect a substantial reduction in sulfur-
oxide levels on a local basis, it is by no means clear that it
will provide the type of broadly-based solutions that will most
certainly be needed in the long run. On this note, we will move
on to examine the second public policy alternative, i.e., economic incentive schemes.

2. Economic Incentives

The basic philosophy inherent in the use of economic incentives for air pollution control is that of a *general reliance on
the allocative mechanisms of the private market, coupled with some
form of exogenous political intervention to correct for specific
deficiencies in the overall system.* Economic incentives might
take several forms including subsidies, fuel taxes, and emission
charges.

Subsidies

Subsidies are intended to encourage the utilization of pollution-abatement schemes by relieving part of the financial burdens that might be imposed on various industries and municipali-
ties. Subsidization can take the form of outright grants-in-aid,
tax or property assessment credits on new pollution control investments, fast depreciation writeoffs, or guaranteed loans. The
perceived need for monetary assistance can be traced to the serious financial plight of some cities (due to a steadily decreasing
tax base) and the unwillingness of many companies to make substantial capital investments that will not increase profits.
Subsidies can be used to encourage or discourage the use of certain abatement techniques, but must be administered wisely so as
to preserve the advantages of flexible decision-making in the individual firm in arriving at a least-cost solution, e.g., stack
gas removal as opposed to fuel substitution.

Subsidization schemes have some specific disadvantages that
make them of limited value. One drawback is that, even with subsidies, most firms may be reluctant to make an investment due to
the absence of any acceptable economic return:

> Thus, if a pollution control device neither helps to
> produce saleable products nor reduces production costs,
> a firm really receives very little incentive to buy the
> device even if the government offers to pay half the
> cost. All that such subsidy schemes accomplish is to
> reduce somewhat the resistance to direct controls.[37]

The accuracy of this statement is acknowledged even within the
business community. One executive has put it this way:[38]

> ...if you would base pollution control on a system of
> incentives, you might be disappointed. The marginal
> dollar gained for pollution control is hardly as ex-
> citing as the marginal dollar gained in expanding sales,
> creating new products or improving technology. This
> type of income promises growth and future profits.
> I think that many, if not most businesses have a short-
> age of key personnel and they would rather use this to
> develop the mainspring of their profits than to maximize
> their pollution subsidies.

Another disadvantage is that outright grants, taken by them-

selves, often discourage development of pollution controls beyond
that which is covered by the subsidy. There is no economic incentive to do more than is possible with the amount of the grant.
Groups may often be reluctant to act until they feel they have
obtained the maximum subsidy possible. This opens the door to
the practice of *gamesmanship* on the part of the subsidized industry. There may be a tendency to overstate the capital needs of
some particular control scheme to increase the subsidy and decrease the share of the burden on the industry itself. Also, in
many cases, there is no clear-cut, end-of-the-line device that is
solely related to controlling pollution. Some modifications in
process might produce valuable by-products whose costs would be
unfairly covered by the subsidy. Unless the government subsidi-
zing body knows the production technologies of all firms involved,
there is no way to make checks on the accuracy of each firm's
cost estimate. Hence, there may be vast informational requirements to be fulfilled at great cost (hiring of experts, etc.)
since information of the kind needed is seldom forthcoming from
the industries themselves!

A third objection to some forms of subsidization can be made
on the grounds that those who benefit from the production of
goods that cause pollution, either by consuming the good or
making profit from its production, should bear the costs of pollution abatement as a factor input to production. The people of
the town where an industry is located should not have to pay the
costs of air and water pollution caused by the industry. Yet
large-scale subsidies, which come directly from tax revenues, are
financed by every taxpaying citizen whether he is involved or
not. There is no preservation of the market function of allocation by price. The subsidy merely transfers the diseconomy from
one group to another, although larger.

Another limitation of subsidization policies is that they
seem suited only for those externalities in which the capital
costs are the only real significant feature that prevents the
situation from being corrected. If effective control technolo-

gies have not been developed for a given pollutant, subsidies
have no real meaning unless applied to a research and development
program.

The case against subsidization as an effective tool in pollution control has been stated very convincingly in a recent
article entitled "Tax Incentives Don't Stop Pollution":[39]

> Federal and State Tax Incentives designed to help in
> the fight against pollution are fiscal carrots that
> don't work. They are expensive, and they are soft on
> pollution. Tax incentives fail because they do not give
> industry an incentive to invest in nonproductive faci-
> lities, they apply only to physical devices, they pro-
> vide the public no gain to offset the revenue loss,
> they are of advantage only to wealthy firms, and they
> shift the burden of reducing pollution to the general
> public.

The conclusion as to the limited usefulness of subsidization
policies applies directly to the area of sulfur-oxide emissions.
There is no really strong evidence to indicate that subsidies
will provide sufficient inducement to power companies and apartment owners alike to invest in new equipment and other pollution
control devices which do not, in general, generate new revenues.
Even with tax credits and direct subsidies, there still remains
a capital expenditure that will yield no return. The only instances in which subsidies seem to be desirable are 1) in helping
states and municipal governments to meet the capital costs of
their own pollution-abatement programs, and 2) in financing basic
research and development. Overall, subsidization schemes are
subject to a number of serious drawbacks that render them inherently unworkable on any broad basis in the control of sulfur-
oxide emissions, or of pollution in general.

Having ruled out the various forms of subsidies as ineffective on any significant scale, we now turn to the second general
class of economic incentives--*direct charges*. These can take the
form of a *fee* on pollutant emissions (determined at the stack) or
a *tax* on the pollutant content of some input to the combustion
process, usually the fuel. In the next subsections we will examine both of these alternatives in depth.

Emission Fees

Emission fees have been suggested as a direct means by which
a polluter is made to come to terms with the costs of external
effects associated with his enterprise. The fee would be levied
in proportion to the amount of effluents discharged. In the
optimal situation, the imposed charge would be equal to the social damages--in whatever way determined--caused by the pollution.
The polluter, now faced with the proper costs of the factor input
of waste disposal, will alter his production methods and/or his
outputs after reevaluating the cost of waste disposal with the
effluent charge attached. The polluter then maintains the flexibility of decentralized decision-making and can use any means
available to arrive at that level of pollution abatement that is
consistent with the objectives of his operation. If the charge
is correctly determined, this level will be one that is consistent with the values of society. He has the choice of 1) cutting
back production to reduce pollution; 2) installing control equipment; 3) changing his process to one that is more efficient pol-
lution-wise, or 4) paying the penalty for the pollution. If the
penalty does indeed reflect the social cost of the pollution, the
latter alternative would only be chosen when society values the
production and consumption of some good more than it does cleaner
air, or when, because of economies of scale, it is cheaper to pay
the penalty to an outside group who will plan, build, and operate
a pollution-control system for multiple users (which may be feasible for water-pollution control, but probably not for air pollution). The big advantage of this scheme is that, if the charge
is chosen correctly, the market mechanisms will lead to an efficient level of pollution, i.e., that level society desires, given
that a certain amount of (least-valued) resources must be expended. Another advantage is that producers are not denied alternative actions on seeking out this least-valued expenditure--
the emission charge provides the stimulation to find innovative
schemes of pollution control even if not for reducing pollution
for its own sake but for profit incentives alone. Obviously, if
similar industries receive like incentives, the one which solves

its pollution problem by minimizing the tradeoff costs of penal-
ties vs. pollution control will reap the largest profit since the
price of the product is established by the market.

Another feature of the emission fee scheme is that part of
the charge would be passed on to the consumer in the form of
price increases and part of the cost would be borne by the producer as part of the cost of doing business. Hence, the results
are equitable since those who benefit from polluting (consumers
and producers) must now pay. This forces a reevaluation of the
benefit they derive from producing/consuming a certain amount of
goods and services. If this benefit is still greater than the
total costs of lost opportunity to society (as now reflected in
the price), then it is to the net benefit of society to have this
amount produced. This leaves room for the fact that there will
be some level of pollution at which society values having other
things more than it values a further reduction in pollution.

Still another benefit of effluent fees is that they can be
levied in proportion to the magnitude of the pollution problems
confronting each different locale. Also, once a metering system
is installed there can be great flexibility in varying the charge
as a function of prevailing meteorological conditions. If these
charges are published in advance, polluters can prepare to switch
to other control means as a temporary measure in certain circumstances. For example, a power plant might find that a stack gas
removal device is the most efficient way for them to control SO_2
emissions under normal conditions and a fixed fee rate. However,
it may be to their advantage to switch to a reserve supply of very
low-sulfur fuel during unfavorable weather conditions when the
stack device may not be the least-cost alternative with an increased emission charge.

If a reliable scheme of emission charges could be implemented,
it would seem to be the ideal solution to the air pollution problem in that it "internalizes" all the external social costs,
thereby preserving the clear advantages of resource allocation by
a properly-functioning market. However, the question of imple-

mentation presents some difficulties that must be resolved before
a workable effluent fee scheme can be designed.

First, there is the problem of correctly determining the
magnitude of the charge or fee so that it accurately reflects the
true social costs involved and distributes them equitably among
the various types of polluters. Harold Wolozin has identified the
measurement of the costs of pollution to individuals and society
at large as one of the most difficult assignments in the economics
of air pollution. He points out that economists "have cast serious doubt on the value and reliability of existing national estimates of damage based on currently accepted definitions,"[40] primarily because of the pervasiveness of the air pollution problem.
While this might appear to pose serious difficulties, we must
keep in mind that we are striving to improve on the present situation, not reach perfection overnight. *A system of fees or
charges has the advantage that it is amenable to trial-and-error
adjustment, is flexible enough to be altered whenever there is
a perceived change in the values (however determined) of society,
and points the system in the right direction (towards optimality
and efficiency) initially.* On these grounds alone, then, it
would be worthwhile instituting effluent charges as an effective
control technique.

A second question that has been raised concerns the nature
of assumptions that are made about investment decisions and business behavior in general. What has been challenged, of course,
is the classical notion of short-term profit maximization, frequently objected to on the grounds that it does not include the
many sociological factors that play important roles in business
decisions and ignores longer-term goals such as the stability or
even survival of the firm.

> ...To support the contention that externalities can be
> internalized through effluent fees, proponents generally
> fall back upon a conventional economic analysis of the
> nature of business behavior in the modern world, a
> model of business behavior which has been questioned
> seriously in the literature on the subject and one which
> very few economists adhere to rigorously in explaining

> the behavior of the firm or industry....Directly related
> to this is the tenuous nature of current theories and
> knowledge about the formulation of investment decisions
> in business firms.
>
> ...Evidence to support the thesis that effluent fees
> will result in investment outlays on pollution abate-
> ment equipment is shaky. Uncertainty, the nature of
> capital markets, and other factors determining invest-
> ment decisions would inject a good deal of indetermi-
> nateness into any attempt to predict responses to
> effluent fees.[41]

While this challenge to the conventional theory of the firm is
well taken, I feel that it does *not* logically imply that effluent
charges are ineffective in bringing about capital outlays on pollution. *If the charge is set high enough, polluters will always
have the incentive to take control measures regardless of their
ultimate objective, be it profit maximization, sales or revenue
maximization, or whatever.* What changes with revised theories
is the exact location of efficient points and not the fact that
we are moving toward efficiency. The crucial assumption behind
effluent fees is that, rather than maximizing profits, producers
will expend the least-valued set of resources to attain a particular objective. Therefore, if a pollution charge is set high
enough, the polluter will always look for a less costly means
of waste disposal, e.g., installing abatement equipment or using
nonpolluting fuels. Also, there is no need to predict accurately
the responses to effluent fees since they could easily be increased
if pollution abatement did not proceed at the desired rate.

The third and, in my opinion, the only *substantive* difficulty with emission fees is in the area of monitoring.

> The real problem which advocates of effluent charges
> must face is the problem of metering, or of estimating
> in some way the amount of effluent actually generated
> by various emitters. Here the problem of air pollution
> is seen to be a particularly difficult one in that the
> number of small emitters and of the number of emitters
> difficult to meter effectively is large and their con-
> tribution to the problem is too great to be ignored![42]

The essence of this problem is in the high cost of existing monitoring devices for large sources and the unavailability of prac-

tical devices for the multitude of smaller sources.

> In-stack instrumentation is already available for mea-
> suring inorganic, gaseous emissions such as carbon di-
> oxide, nitrogen oxides, and sulfur oxides resulting from
> fossil-fuel combustion. But before the gas from a stack
> can be sampled, expensive scaffolding and a platform
> must be built on the stack and probe holes provided.[43]

While metering costs may be prohibitive at present, the technology is undergoing development that could produce economical devices in the very near future. Also, what we consider to be
"economical" today could be vastly different in a couple of years
as the problems of pollution continue to degrade the human environment. Still, possible difficulties in inspection, measurement and administration of what might be a complex system are the
greatest obstacles to an effective effluent fee scheme. But the
benefits of such a system are potentially great enough to warrant
a careful economic evaluation of the costs that would be incurred.
Unfortunately, such analyses seldom seem to be done for reasons
that have been described by Marshall Goldman:

> The first hurdle that must be overcome is the winning
> of political support from the numerous skeptics who
> doubt that economic controls are workable. Many pollu-
> ters distrust the use of economic controls. Some dis-
> trust them because they do not understand them. Others
> cite the fact that economic controls have not always
> worked well. Occasionally arbitrarily applied taxes
> and subsidies have solved one set of problems only to
> create a whole new set of distortions. Thus some
> critics fear that the use of pollution charges will
> bring about just the opposite of what is intended.[44]

If some of these attitudes can be overcome, a system of effluent
fees seems to have great potential as a long-run solution to the
overall problems of pollution.

Fuel Taxes

A good example of the second form of economic incentive is
the imposition of a tax on high-sulfur fuels in such a way that
the cost of producing them becomes comparable to the cost of low-
sulfur, less-polluting fuels. Such a tax is, in one respect,
another form of an emission charge since, given the known sulfur

content of fuel and the efficiency of a particular process,
the level of emissions is directly related to the amount of
fuel burned. But, since the tax is on the *input* to the pollution
process, it might be much easier to administer since all that
needs to be known to determine the charge is the amount of fuel
consumed. While the problem of determining the correct rate
structure still remains, again it is more one of making sure the
incentive for polluters to take action is strong enough rather
than assuring that the tax accurately reflects the true costs
to society of pollution. When the concentrations of sulfur
oxides are a good deal above the minimum acceptable levels,
we are not so much concerned with an optimal solution (in the
long run) as we are with doing something about the problem on
a more immediate basis.

A tax of this sort, if imposed on the producers of the various fuels, has the advantage that it would provide an incentive
for those producers to develop less costly ways of producing low-
sulfur fuel since the demand for this type would increase. A
progressive (over time) rate structure could give the industries
a chance to make these advances through accelerated technological
research. Much of the tax would be passed on to the users of
high-sulfur fuel in the form of higher fuel prices. If the tax is
properly determined, the cost of burning high-sulfur fuel will
not be attractive compared to the cost of its low-sulfur counterpart. Thus, there will be a greater demand for the low-sulfur
fuel, providing an incentive for producers to gear up production
facilities to meet the demand. The increased costs would be
spread out down the line from producer to the ultimate consumer,
which seems to be the equitable distribution of the pollution
burden.

One serious disadvantage of this scheme as it now stands
is that there is no incentive to users of the fuel to develop
other means of sulfur-oxide control, e.g., through stack gas
removal technology. To counteract this, it has been suggested
that a tax rebate be available to encourage the users to find

alternative, least-cost methods to reduce their emissions. Thus, there appears to be a desirable double incentive--to the producers to clean their fuel before sale, and to the users to take alternative measures to reduce their emissions. We can see how a scheme such as this might lead to a least-cost technological solution. If fuel users find that they can develop ways of burning high-sulfur fuel (by taking the sulfur from the stack gases) that is less costly than paying the price for low-sulfur fuel, they will go ahead with it and demand more high-sulfur fuel. In this case, the technology of stack gas removal would be less costly than that of desulfurization, and the most efficient result would be continued production of high-sulfur fuel.

This proposal is similar in some ways to both direct regulation and effluent fee schemes. As with direct regulation, it focuses attention on one particular control means (making low-sulfur fuel more attractive) and allows for the development of other techniques via a tax rebate. As with effluent fees, on the other hand, it incorporates a variable rate structure and attempts to "internalize" the economic costs of polluting the air (rather than simply outlawing the use of high-sulfur fuel). The question that must be asked is: does this proposal successfully bypass the difficulties with effluent fee schemes (monitoring, especially) without incurring the problems associated with direct regulation?

The answer to me seems to be *no*. It may be true that a properly-formulated legislative package can avoid the problem of continuous monitoring and succeed in retaining the flexibility of decentralized decisions, encouraging the development of a wide range of technologies, and seeking an efficient level of pollution through market mechanisms. *But the prohibitive drawbacks of direct regulation remain in the long-run.* For such a system to work, regulating agencies must still have swift enforcement with practical detection and measurement techniques to uncover violators. Without this, there is no way to assure compliance.

Once a polluter secures a variance (and a tax rebate) for a
new device he has no incentive to keep that device operating
efficiently unless violations can be detected and punished.
Another very important shortcoming is that a tax of this sort,
especially at the federal level, is inflexible with respect
to varying geographic locations and meteorological conditions.
The danger here is that state and local governments will grow
to rely on a federal taxing scheme without gearing up to meet
longer-term issues, since serious questions will be raised ultimately as to the value of reducing pollution beyond a certain
level.

Most of the other objections to direct regulation apply
equally well to the taxation scheme. The only real difference
between the two policies is that one imposes a variable tax on
sulfur content while the other simply outlaws it beyond a certain
level. In fact, if the tax were high enough, the two schemes
would be virtually identical. So it is not by any means clear
that a fuel tax can be a successful "marriage" of the advantages
of economic incentives and direct regulation that avoids the
associated disadvantages.

3. Concluding Remarks

Due to the tremendous complexities of the problems of pollution, no one policy alternative can at present solve a given
problem by itself, since each has some very serious drawbacks.
However, a well-chosen mix of policy tools, making use of the
best characteristics of each while taking steps to counteract
associated disadvantages (or making a value judgment as to the
least of many evils), seems to be the best approach to the solution of pollution problems. Our final task in the remaining section is to try to establish some general guidelines for the design of regional and national pollution-abatement policies. We
will look at the roles of the federal, state, and local governments in controlling sulfur oxides, taking an overview of all the
economic, technological, and policy-making factors considered
so far.

VI.   CONTROLLING SULFUR OXIDES: AN OVERVIEW

Initially, governmental response to the growing problem of
sulfur oxide air pollution came at the state and local levels.
Under the pressures of public opinion to take immediate action,
the cheapest and most politically feasible alternative has been
direct regulation, usually in the form of a restriction on the
sulfur content of fuels. Harold Wolozin has commented on this
trend:[45]

> ...With the emphasis placed by the Air Quality Act of
> 1967 on governmental enforcement of air standards and
> the designating of air quality control regions, the
> trend seems to be toward increasing governmental as-
> sumption of direct responsibility rather than any
> commitment to the indirect pressures operating through
> market incentives such as effluent fees. In a sense,
> the conceptual battle lines have been drawn.

Based on the discussions of the previous section, I feel that this
may be a dangerous trend that risks delaying any really effective
abatement of sulfur oxide pollutants. While direct regulation
may have some immediate short-run advantages, its effects in the
longer run may well be counterproductive. First of all, restrictions on the use of high-sulfur fuel provide no incentive for
polluters to search out least-costly abatement schemes, and when
provision is made for variances, there is no incentive for abatement levels to be maintained unless there is a workable detection
and measurement scheme. Thus the argument about the primitive
state of the art of measurement technology is just as much a drawback to direct regulation as it is to an economic incentive
scheme. Secondly, direct regulation is usually inflexible with
respect to changing meteorological conditions, a serious drawback
as overall pollution levels begin to decrease and justification
for further controls becomes more difficult to make. This is just
another of the many ways in which direct regulation can work to
discourage movement in the direction of economic efficiency. A
third and perhaps most damaging effect of direct regulation is
the large amount of additional uncertainty that it introduces into
the economic arena. I have already speculated on the reluctance

that an oil company might have to respond to short-term changes
in the demand for low-sulfur oil due to a rash of local regulations,
many of which could be revoked at any time in the future. A recent article in _Time_ seems to bear this out:[46]

> ...Last week at General Motors' annual meeting, Chairman
> James Roche announced that the corporation will spend
> $214 million to combat pollution in 1971. Despite these
> outlays, environmentalists charge that major polluters
> often stall for time during lengthy negotiation periods
> provided in many state and local laws, then begin work
> in earnest only when court action is threatened. In re-
> plying to this criticism, industry executives note that
> there are still no nationwide standards for many kinds
> of pollution control. If federal laws become tougher
> than local ones, they note, much of their early invest-
> ment could be wasted. Says Crown-Zellerbach President
> C. R. Dahl: "Standards have a way of changing on us, we
> never really know where we will be tomorrow."

Based on the above considerations, I feel that direct regulation is a generally inadequate policy tool when applied to the
problem of controlling sulfur oxide emissions. While it may be
the quickest, cheapest, and most politically feasible means to
abate pollution on a local level, it is basically ineffective as
a long-term solution to the problems and risks creating a dangerous legislative momentum that will be counterproductive in the
long-run. However, regulation may be effective in supplementing
other schemes when no feasible policy alternatives exist to correct for specific deficiencies.

2. The Alternative to Regulation

Recognizing the basic inadequacy of direct regulation as a
pollution control scheme, we should divert our attention to the
alternative category--economic incentives. A scheme of this sort
is presently being considered at the federal level where recent
proposals[47] have been made regarding a tax on the sulfur content
of fuels. In a recent communication,[48] Gordon J. F. MacDonald of
the Council on Environmental Quality has expressed the rationale
for moving in the direction of financial incentives:

> The factors of a changing technology combined with the
> schedule and enforcement of air quality standards pre-

sage delay until industry installs the needed equipment. The introduction of a financial incentive will supply a strong motivation for them to actively pursue the most efficient techniques or combination of methods for reducing emissions with an incentive to effectively maintain and operate facilities as well as to install them. The liability for payment of a charge on uncontrolled emissions would produce the incentive to bring about the quickest possible reduction in the emission of this harmful pollutant (SO_2) by making profits depend directly upon the degree of control undertaken.

In this context, the issue at hand becomes one of deciding among the various forms that economic incentives might take. I think that subsidization schemes such as outright grants or tax incentives can be eliminated from serious consideration on the basis of previous discussions. That leaves us with emission charges and fuel taxes as alternatives that must be viewed within the overall context of the sulfur oxide problem. A brief review of that context will be useful at this point.

Some of the key facts brought out in previous discussions are as follows:

1. the technology for removing sulfur from stack gases is undergoing intensive development but is not yet feasible for widespread application

2. in general, stack gas technology has potential usefulness only for large sources because of the high capital costs involved

3. the supply of low-sulfur oil can be considered adequate both in the short-term and long-term since it depends only on desulfurization technology, which is well-developed

4. the availability of low-sulfur oil is a function of demand at a premium price

5. the supply of low-sulfur coal is abundant in the short-term but limited in the long-run relative to the supply of high-sulfur reserves

6. desulfurization technology for coal, though under development, is not presently available for widespread application

7. the availability of low-sulfur coal is a function of demand at a premium price

8.   the instrumentation necessary to continuously moni-
     tor the emissions of both large and small sources
     is not available for widespread application at pres-
     ent; again the costs are prohibitive for the smaller
     sources

Two things are important to note from this review. First, the
number of feasible abatement techniques is strongly a function of
time. At present, the use of low-sulfur fuels is generally the
only realistic choice. However, as other pollution-control technologies are developed, this situation will become very different in the long-run. This disparity between short- and long-
term availability of technology applies as well to instrumentation for detection and measurement. Secondly, large and small
pollution sources are in distinctly different situations with regard to the present and future availability of technical alternatives for sulfur oxide control. Most research and development
is channeled in the direction of devices for the large sources
since the relative costs of effective control devices for the
smaller ones are very prohibitive. Thus it appears that the use
of low-sulfur fuel will remain the only practical alternative for
small polluters for a much longer time than for the larger
polluters.

In our discussions of the economic incentive policies that
might be employed to control pollution, we have noted disparities
similar to the ones mentioned above regarding effects over the
short- and long-term and in relation to the size of the pollution
source. For example, fuel taxes seem useful in the short-term,
but are inflexible to changing meteorological conditions, an important issue in any effective long-term abatement scheme. Alternatively, the concept of emission fees seems best suited to
the long-term while the needed monitoring instrumentation is not
at present available, especially for the multitude of small
sources. This suggests that the best pollution control policy
is one which matches the advantages of different schemes to the
changing situations encountered (1) over time and (2) between various pollution sources. By matching certain policy tools to certain segments of the overall problem in a *staged strategy over*

time, we can preserve a great deal of flexibility in seeking
efficiency, avoid dangerous legislative momentum that can per-
petuate a given policy beyond its usefulness, and bring about
immediate and substantial reductions in sulfur oxide levels.
How then do we go about designing such a strategy that can effectively come to grips with the problem of controlling sulfur oxide
emissions?

2. <u>Concluding</u> <u>Remarks</u>: <u>A</u> <u>Proposal</u>

I envision an approach that will ultimately internalize the
costs of pollution through a *two-phased* strategy that uses economic charges as the fundamental policy tool:

Phase One

...A progressive tax on the sulfur content of fuel should
be enacted at the federal level with provision for rebates
for alternate abatement techniques (even though few are
feasible at present)

...The revenue generated should be used to subsidize large
scale research and development efforts and assist the
states and cities with the financial burdens of controlling municipal pollution activities

...Direct regulations should be phased out immediately at
the local levels. If pollution abatement does not proceed satisfactorily under the federal tax, the states
should enact supplementary taxes until the total charge
is sufficiently high to effect the required abatement.
Direct regulations should be used only as a last resort.

Phase Two

...Effluent fee schemes at the state and local levels
should ultimately replace the federal fuel tax, beginning
with larger sources and then smaller sources if possible

...If necessary, a two-fold policy of emission fees for
large sources and a fuel tax for small sources could be
employed until a feasible monitoring scheme for small
sources is available *or* if low-sulfur fuel appears to be
the only control alternative for the small sources in the
long-run

...Ultimately the federal fuel tax will be discarded in
favor of state or local emission fees and fuel taxes as
revenue requirements switch from the national (basic re-

search) to the local (administrative) levels

This proposal reflects the preference ordering established in previous discussions with regard to alternate policy tools, i.e., emission fees are the most desirable ultimately, followed by fuel taxes as a suitable interim alternative, with direct regulation a poor third to be phased out as soon as possible. The approach has some particularly attractive features. With emission fee schemes generally not feasible at present, a fuel tax has a number of advantages over direct regulation in the short-term, even though this scheme suffers the same drawbacks as direct regulation in the long-run. The most significant benefits of the tax are as follows:

1. It provides a more uniform set of regulations that reduce somewhat the uncertainties introduced by legislative dabbling in the economic arena at a multitude of local levels

2. It generates revenue that can be applied to needed research and development efforts, which can most effectively be administered at the federal level

3. It can provide an immediate and strong incentive to reduce the levels of pollution

4. It avoids the problem of detecting and monitoring emissions as long as no sufficiently attractive alternatives to low-sulfur fuel exist. Since the effectiveness of the tax does not depend on technology right away, it allows time for the parallel development of abatement and measurement devices

5. In temporary situations where the use of high sulfur fuel is necessary, the polluter pays a direct and high penalty whereas under regulation he may be willing to break the law and risk court action. An automatic charge provides much more incentive than the possibility of a fine, especially when there is the chance that a violation will go undetected under direct regulation

6. The low-sulfur fuels whose use is encouraged are readily available on the short-term

While this alternative seems best suited to present needs, the shortcomings that it shares with direct regulation become more serious as technical development progresses. Unless a workable

detection scheme is available, compliance with standards will become less certain as more variances are granted to alternative
control techniques. In addition, the varying local effects of
meteorological conditions and geographic location will become
increasingly important as overall pollution levels decrease. It
is the basic inflexibility of the fuel tax in adjusting to these
changing factors that make it unsuitable as a long range solution. For this reason we turn to the emission fee scheme in
Phase Two.

We have seen that if the problem of monitoring can be overcome, emission charges have the advantage of maintaining the
greatest flexibility in achieving economic efficiency with the
least amount of collective interference in the economic arena.
The staged strategy is advantageous in this regard since it allows time for the development of the required technology while
still doing something significant to control the problem in the
meantime. The result is ultimate reliance on the most desirable
long-term solution--emission fees--while avoiding present problems
of implementation and taking advantage of the shorter-term benefits of the fuel tax.

In addition to bringing about a "marriage" of the short- and
long-term advantages of different policy tools, the staged strategy over time has the following beneficial side effects:

1. It reduces the uncertainties that are inevitably
    introduced through legislative dabbling in the
    economic arena. A progressive fuel tax provides
    uniformity when established at the federal level, and
    an emission charge does not involve regulation of the
    use of fuels or abatement devices.

2. By recognizing a disaggregation of the effects of
    policies over time, it maintains the incentive neces-
    sary for the continued development of new technologies.

3. It allows for the use of a mixture of policy tools,
    both in the short- and long-terms, which helps avoid
    the legislative "momentum" that risks foreclosure of
    important options in the future.

All of these considerations lend added weight to our fundamental conclusion in this analysis, i.e., the most effective

policy regarding the control of sulfur oxide emissions appears to
be a combination of fuel taxes and emission fees in a staged
strategy over time. If properly formulated, such a policy can
realize the efficiency-seeking advantages of each scheme while
avoiding the shortcomings that make sole reliance on either one
unrealistic. At the same time, there are additional benefits in
adopting a policy that anticipates changing situations and maintains the flexibility needed to deal with them as they materialize.
Thus, the proposal presented here adheres to the principal guide-
line that we have recognized with regard to pollution-control
policies--the preservation of the efficiency-seeking mechanisms
of decentralization and flexibility--and is sensitive to the real-
world context within which implementation must take place.

REFERENCES

1. Council on Environmental Quality, <u>Environmental Quality</u>, The First Annual Report, August 1970, p. 5.

2. U.S. Department of Health, Education, and Welfare, "Air Quality Criteria for Sulfur Oxides," Washington, D.C., January, 1969, p. 2:19.

3. <u>Ibid</u>., p. 1:26.

4. Conference on the Study of Critical Environmental Problems, <u>Man's Impact on the Global Environment</u>, MIT Press (1970), p. 13.

5. U.S. Department of Health, Education, and Welfare, "Air Quality Criteria for Particles," Washington, D.C.

6. See Reference 2, p. 10:21.

7. <u>Ibid</u>.

8. <u>Ibid</u>., p. 10:22.

9. <u>Ibid</u>., p. 7:23.

10. The American Chemical Society, <u>Cleaning Our Environment - The Chemical Basis for Action</u>, Washington, D.C. (1969), pp. 76-77.

11. <u>Ibid</u>.

12. See Reference 2, p. 10:8.

13. <u>Ibid</u>., pp. 9:61-62.

14. <u>Ibid</u>., pp. 9:62-64.

15. <u>Ibid</u>., p. 10:19.

16. <u>Ibid</u>., p. 10:22.

17. See Reference 10, p. 81.

18. See Reference 2, pp. 2:5-6.

19. Thomas K. Sherwood, "Must We Breathe Sulfur Oxides," <u>Technology Review</u>, January, 1970.

20. U.S. Department of Health, Education, and Welfare, "Control Techniques for Sulfur Oxide Air Pollution," Washington, D.C. (1969), p. xxxi.

21. <u>Ibid</u>., p. xxxii.

22. See Reference 19.

23. See Reference 20, pp. xxxii-xxxiii.

24. See Reference 19.

25. Ibid.

26. See Reference 20, p. xxviii.

27. See Reference 19.

28. See Reference 20, p. 4:135.

29. See Reference 19.

30. See Reference 20, p. 4:29.

31. Ibid., p. xxx.

32. See Reference 2, p. 2:5.

33. Arthur M. Squires, "Clean Power from Coal," Science,
     August 28, 1971, pp. 826-828.

34. Lawrence W. Pollack, "Legal Boundaries of Air Pollution
     Control - State and Local Legislative Purpose and Techniques,"
     in Duke University School of Law, Law and Contemporary
     Problems, Vol. XXXIII, No. 2, Spring 1968, p. 343.

35. Allen V. Kneese, "How Much is Air Pollution Costing Us in
     the United States," in Proceedings: The Third National
     Conference on Air Pollution, p. 539 (Public Health Service
     Pub. No. 1649, 1967).

36. See Reference 34, p. 354.

37. Edwin S. Mills, "Economic Incentives in Air-Pollution Con-
     trol," in H. Wolozin, ed., The Economics of Air Pollution,
     Norton & Co. (1966), pp. 44-45.

38. Harold Wolozin, "The Economics of Air Pollution: Central
     Problems," in Duke University School of Law, Law and Contem-
     porary Problems, Vol. XXXIII, No. 2, Spring 1968, p. 236.

39. Arnold W. and Glenn Reitze, "Tax Incentives Don't Stop Pollu-
     tion," American Bar Association Journal, Vol. 57, February,
     1971, p. 127.

40. See Reference 38, p. 228.

41. Ibid., pp. 235-236.

42. Vickrey, "Theoretical and Practical Possibilities and Limitations of a Market Mechanism Approach to Air Pollution Control," p. 5 (presented at annual meeting of the Air Pollution Control Association, Cleveland, Ohio, June 11, 1967).

43. John H. Ludwig, "Air Pollution Control Technology: Research and Development on New and Improved Systems," in Duke University School of Law, Law and Contemporary Problems, Vol. XXXIII, No. 2, Spring, 1968, p. 225.

44. Marshall Goldman, "Pollution: The Mess Around Us," in Controlling Pollution, Prentice-Hall, Inc. (1967), p. 36.

45. See Reference 38, p. 233.

46. "What the Pollution Fight Will Cost Business," Time, May 31, 1971, p. 82.

47. "Nixon Seeks Sulfur Tax as Antipollution Incentive," New York Times, February 2, 1971.

48. Personal communication with Gordon J. F. MacDonald in a letter dated March 10, 1971.

CHAPTER 5

WATER QUALITY IMPROVEMENT
IN BOSTON HARBOR

by

Dennis Ducsik
Thomas Najarian

ABSTRACT

The utilization of Boston Harbor and its island group to their full potential as a recreational resource hinges on one important condition--that of an acceptable water quality. Unfortunately, the present status of pollution in the harbor prohibits the effective development of water-related recreational facilities on any broad scale. This article is intended to point out an area in which feasible abatement measures have the greatest potential for bringing about a pronounced water quality improvement. This involves the issue of sludge handling and disposal.

The current method of sludge disposal at the Boston treatment plants is to dump it, after bacterial digestion, directly into the harbor. Yet it is known that the disposal of digested sludge through direct discharge into receiving waters greatly reduces the overall effectiveness of the treatment plant in removing bacteria and oxygen-demanding materials, and considerably negates whatever nutrient removal there might otherwise be. In investigating the case against sludge, we have found substantial evidence indicating that sludge is a primary contributing factor to the bacterial degradation of the waters in Boston Harbor.

As a first step in finding a suitable disposal scheme for digested sludge which avoids harbor dumping, we have examined in a preliminary way some commonly-used techniques. One feasible method is drying and storage on land. This analysis was not intended to be complete--we realize that there may be other more efficient ways to handle sludge. Our primary purpose has been to focus attention on the immediate need to attach a high priority to the entire question of sludge and its effect on the quality of receiving waters.

WATER QUALITY IMPROVEMENT IN BOSTON HARBOR

I.   INTRODUCTION

We have seen in Chapter 3 that shoreline recreation is a
public good and, as such, will be allocated an inefficiently
small share of coastal acreage by the private market. As a
short-term component to the solution of this problem we recommended that governments take immediate steps to preserve whatever
suitable areas remain, especially in urban regions where the
demands are greatest. One such area is Boston Harbor, where
a large number of undeveloped islands offer a unique opportunity
to provide facilities to meet the future recreational needs
of the Boston metropolitan region. However, the value of the
harbor islands as a recreational resource hinges on one important
condition, that of an *acceptable water quality*:

> The Metropolitan Area Planning Council has recently com-
> pleted an open space and recreation study of Boston Harbor.
> The Council considers the harbor a major recreational
> center for the Boston area and recommends a program of
> open space acquisition and development....The MAPC, how-
> ever, points out, "No improvement or recreational devel-
> opment of the harbor is possible without an end to
> pollution.[1]

Unfortunately, the present status of water quality in much of
Boston Harbor prohibits the effective development of water-
related recreational facilities. The purpose of this chapter is
to seek out and discuss an area in which feasible abatement controls have the greatest potential for bringing about a pronounced
improvement in water quality in the harbor area.

II. BACKGROUND

Boston Harbor consists of an inner and an outer harbor, and
a number of bays, as can be seen in the aerial photograph of
Figure 5.1. The Sierra Club has described the harbor area in a
fact sheet prepared by a task force of their Eastern New England
Group:[2]

Source: Aerial Photos of New England, Boston, Mass.

Figure 5.1   Boston Harbor and Vicinity
             (Cape Cod in Background)

It is an understatement to say that metropolitan Boston
has failed to make imaginative use of its unique resource.
A Civil War vintage prison denies the public any access
to Deer Island. A city hospital for the chronically ill
and infirm on Long Island accomplishes the same. For
years Spectacle Island was a city dump, and today the
Boston Redevelopment Authority (B.R.A.) moors barges of
burning refuse from urban renewal off the Brewsters, the
Harbor's most scenic islands.

Elsewhere, highways built atop the waterline and houses
built to the shoreline obstruct access. Oil tank farms,
shopping center parking lots, warehouses, and other commercial uses are taking over more of shoreline and marshes
with a blatant disregard for the uniqueness of their locations.

Pollution is, of course, a major problem. Logan Airport
fouls the air and creates noise. These add to the problems caused by the vast quantities of raw and treated
sewage, and some industrial waste, which are discharged
into the Harbor to be carried away by the tides. Sewage
and oil spills cause the closing of several public beaches
each summer.

Many of the Harbor's rich clam beds are closed to the public. Some are open only to commercial rakers who must
clean and treat their take before marketing. Clamming is
potentially an industry as well as a substantial recreation resource.

Fill has been so extensive that the Inner Harbor is today
little more than a corridor to Boston's wharves, an open
sewer through which the polluted waters of the Charles and
Mystic Rivers and much of the City's wastes can reach the
sea.

Boston Harbor is much more, however, than fill, airport
noise, and water pollution. It is big--47 square miles
of water, 180 miles of tidal shoreline, 30 islands with
a total area of about 1400 acres. While the Inner Harbor
is congested and dirty, the Outer Harbor has a lot of open
water. There are islands with trees and open meadows, some
with rocky shores, many with mysterious ruins and old forti-
fications....Those marshes that remain are a vital link
in the ecological chain that supports marine life far from
the limits of Boston Harbor's waters. They have a monetary
value that planners frequently forget.

Despite very poor access, the Harbor's public beaches are
used by about 1.5 million bathers during a summer season.
There are 28 boat launching facilities and 35 yacht clubs
with more than 3,000 member families.

The Massachusetts Legislature has recognized the unique importance of Boston Harbor and its islands by authorizing the purchase of those islands that are now in private hands.[3] The islands are to be used for public recreation and conservation. At the same time, the New England River Basins Commission has established water quality goals and has been engaged in the development of a water quality management plan for the Boston Harbor Drainage Area.[4]

It is within this context that we examine the problem of pollution in Boston Harbor. This chapter is intended to serve as an input to the process of managing and planning for the effective abatement of water pollution in the harbor. Our approach will be to focus on a specific aspect of the problem with the goal of determining the most feasible course of action that will make a substantial contribution to improvement of the harbor water quality. The results of this analysis are presented in the following sections.

III. POLLUTION IN THE HARBOR--AN OVERVIEW

1. The Status of Harbor Pollution

In Massachusetts there are three basic categories of water quality, as shown in Table 5.1. Class SA waters are the cleanest and are suitable for all forms of recreational activity. Class SB waters are deemed suitable for bathing and restricted shellfishing, but are tolerable only in a marginal sense since some people might prefer to avoid contact with water in this class. Class SC waters are not suitable for water-contact activities or shellfishing, but can be used for boating.

These classifications are based on allowable concentrations of a number of indicators including dissolved oxygen, coliform bacteria, and plant nutrients such as dissolved phosphorus and nitrogen. *Dissolved oxygen* is necessary to sustain fish and other marine life and is depleted in the biochemical decomposition of organic matter in sewage or by an overabundance of oxygen-demanding plant life. The presence of *coliform bacteria* is

Classification	Usage
SA	Suitable for any high quality water use including bathing and water contact sports. Suitable for approved shellfishing areas.
SB	Suitable for bathing and recreational purposes including water contact sports; industrial cooling; excellent fish habitat; good esthetic value and suitable for certain shell fisheries with depuration.
SC	Suitable for esthetic enjoyment; for recreational boating; habitat for wildlife and common food and game fishes indigenous to the region; industrial cooling and process use.

Source: Commonwealth of Massachusetts, Water Resources Commission

Table 5.1 Commonwealth of Massachusetts Water Quality Classifications and Usage

indicative of the existence in the waters of pathogenic bacteria which constitute a health hazard:

> ...Ingestion of these pathogens by drinking polluted water or by eating raw or partially cooked shellfish grown in these waters can cause gastrointestinal diseases such as typhoid fever, dysentery and diarrhea. The infectious hepatitis virus, as well as other enteric viruses, may also be present. Body contact with water polluted by bacteria can also cause eye, ear, nose, throat or skin infections. Therefore bacterial pollution presents a health hazard, not only to those who come in contact with polluted waters, but also to those who may eat shellfish taken from the waters.[5]

Finally, *nutrients* in the water provide a food source for plants and phytoplankton which, when overly abundant, can seriously reduce the dissolved oxygen content of the water. The Massachusetts water quality classifications as a function of these and other primary indicators are shown in Table 5.2. These classifications provide the basis for the water quality goals that have been established for Boston Harbor by the Massachusetts Division of

	CLASS SA	CLASS SB	CLASS SC
Coliform bacteria (per 100 ml)	Not to exceed a median value of 70 and not more than 10% of the samples shall ordinarily exceed 230 during any monthly period	Not to exceed a median value of 700 and not more than 10% of the samples shall ordinarily exceed 2300 during any monthly period	None in such concentrations that would impair any usages specifically assigned to this class
Sludge deposits-- solid refuse, floating solids, oil, grease, scum	None allowable	None allowable	None except that amount that may result from a waste treatment facility with appropriate treatment
Dissolved oxygen (ml)	Not less than 6.5 at any time	Not less than 5.0 at any time	Not less than 3.0 at any time. Not less than 5.0 during at least 16 hrs of any 24-hr period
Total phosphate (ml)	Not to exceed an average of 0.07 as P during any monthly sampling period		
Ammonia nitrogen (ml)	Not to exceed an average of 0.2 as N for any monthly sampling period		Not to exceed an average of 1.0 as N for any monthly sampling period

Note: In addition to the above standards, the waters shall be substantially free of pollutants that will 1) unduly affect the composition or physical or chemical nature of bottom fauna; 2) interfere with the spawning of fish or their eggs.

Source: Commonwealth of Massachusetts, Water Resources Commission, "Water Quality Standards," 1968.

Table 5.2   Massachusetts Water Quality Standards

Water Pollution Control, as shown in Figure 5.2.

As of the summer of 1967, the actual water classifications
and average coliform bacteria counts in the harbor were as shown
in Figure 5.3. Comparison of this information with the water
quality standards for bacteria indicates that the harbor had been
grossly polluted. Water quality tests conducted in 1967 by the
Federal Water Pollution Control Administration yielded the following results:

- With regard to *coliform bacteria*:

  > ...excessive counts of coliform were found. Total
  > coliform counts as great as 520,000 per 100 ml were
  > found in the Inner Harbor Area. In general, very
  > high counts were found in the northern portions of
  > the harbor, while Quincy, Hingham and Hull Bays in
  > the southern portion would probably meet Class SB
  > water quality criteria for bacteria...[6]

- With regard to *dissolved oxygen*:

  > ...of the eighteen stations sampled during July and
  > August of 1967, only six met the Class SC standard.
  > Furthermore, only two stations met the tentative
  > recommendations of the National Technical Advisory
  > Committee, that "Dissolved oxygen concentrations in
  > estuaries and tidal tributaries shall not be less than
  > 4.0 mg/l, at any time or place...for protection of
  > marine resources."....Excessive phytoplankton acti-
  > vity is suggested by the wide fluctuation of dissolved
  > oxygen during the latter portion of the 1967 survey.[7]

- With regard to pollution by oxygen-demanding, *organic
  matter* (primarily in sludge):

  > ...All reaches of Boston Harbor and each of its
  > tributary streams, except the inland marine reaches
  > of the Weir and Weymouth Back Rivers, were polluted.
  > Based upon the biological conditions about seven
  > square miles, or 30 percent of the Harbor, were
  > grossly polluted. Chemical analysis of harbor sedi-
  > ments for carbon and nitrogen support the biological
  > findings of organic enrichment. Extensive deposits,
  > some greater than three feet deep, of decaying organic
  > matter and incorporated oily residues covered much
  > of the Harbor.[8]

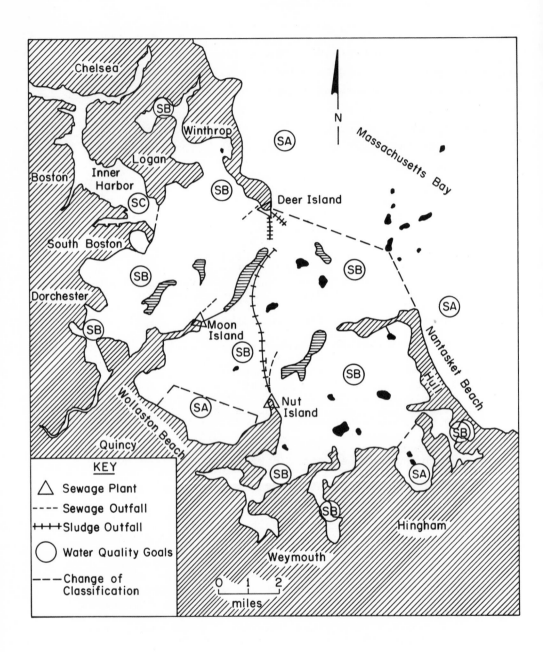

Figure 5.2   Water Quality Goals for Boston Harbor

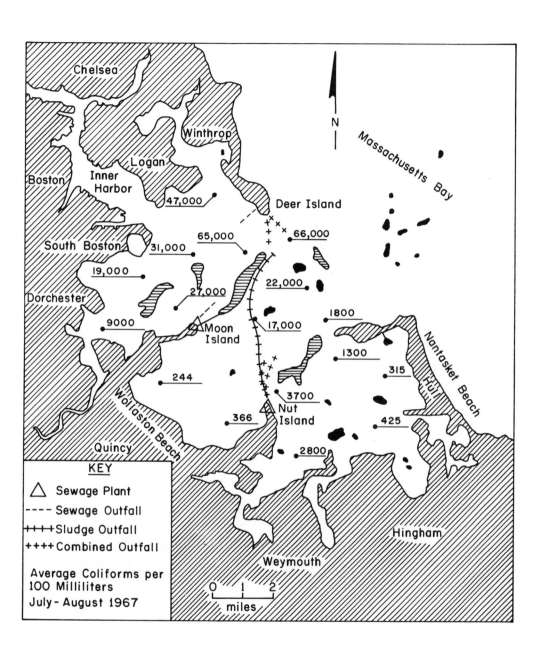

Source: See Reference 1.

Figure 5.3   Bacterial Pollution Densities--Summer, 1967

- With regard to *nutrients*:

> ...The average values of ammonia nitrogen and soluble
> phosphorous were equal to or greater than 100 and 40
> micrograms per liter, respectively, in all areas of
> Boston Harbor inland from its mouth near Masschusetts
> Bay. Such high concentrations of nutrients caused
> overly enriched conditions that stimulated dense
> populations of phytoplankton which exceeded 1,000 per
> milliliter in about sixteen square miles, or 66 percent
> of the Harbor....In addition to causing excessive
> phytoplankton populations, the nutrients stimulated
> dense growths of attached marine plants...on most
> buoys, piers, and marine facilities. Several intertidal
> and shallow areas of the harbor and certain reaches of
> Winthrop Bay supported dense growths of attached marine
> algae. These caused noxious odors in Winthrop Bay,
> unsightly growths at marine facilities and increased
> maintenance costs associated with buoys and piers.
> In Winthrop Bay, decomposing masses of sea lettuce
> have caused hydrogen sulfide emissions sufficient to
> discolor paint on nearby dwellings.[9]

As a direct result of this high degree of pollution in the harbor,
by April of 1968 60 percent of the shellfishing acres had been
closed, while another 29 percent were restricted. In addition,
many city beaches had been closed as a result of the health
hazard presented by the high pathogenic bacteria levels. Boston
Harbor, at this point in time, was little more than a cesspool,
serving as the terminus of the Boston sewer system. Raw sewage
was being discharged at both Deer Island and Moon Island as well
as by combined storm and sewer overflows.

> Sewage-like solids, other assorted rejectamenta, and
> oily slicks also were observed in the surface waters of
> most portions of Boston Harbor. Such materials were
> abundant near the Deer Island sewer outfalls at the
> mouth of Boston Harbor, near Moon Island, the north end
> of Long Island (Nut Island sludge outfall) and the in-
> land reach of Quincy Bay.[10]

The condition of the harbor was, in short, extraordinarily bad!

In May of 1968, the long-awaited Deer Island treament plant
went into operation. This had been seen as a major weapon in the
fight against harbor pollution, as indicated by the following
headlines in a local newspaper in December of 1967:

HARBOR POLLUTION: A RAY OF HOPE

> Every day, Boston and the MDC still pump 350 million
> gallons of raw sewage into Boston Harbor. But it looks
> now as though this health threat may be eliminated as
> early as next summer.[11]

Unfortunately, such optimism proved to be premature. During the
summer of 1968, further water quality tests by the Federal Water
Pollution Control Administration[12] indicated that, while there
were some notable improvements in the coliform counts in some
areas, the majority of the harbor (particularly the outer portions) was still severely polluted. The average coliform counts
for the water in this period are shown in Figure 5.4. Comparison
with the 1967 results in Figure 5.3 show that little or no net
improvement had taken place:

> ...The average coliform densities during the summer of
> 1968 were about the same or, in some areas, significantly
> more than the preceding summer. Quincy Bay, however,
> showed an improvement in water quality over 1967, and
> met the standard for "SA" classification...insofar as
> the coliform density is concerned.

> ...The waters adjacent to the outfalls of Deer Island
> may be described as polluted even when the sewage efflu-
> ent was chlorinated. This unquestionably was due to the
> limited chlorination capacity of the Deer Island and Nut
> Island facilities.[13]

This data clearly indicates that, while the water quality remained
marginally good for recreational purposes in some bays and town
harbors, *no real progress* had been made towards bringing about
the stated water-quality goals. Figure 5.5 shows that the northern and central portions of the harbor continued to violate the
standards approved by the state. Yet, as can be seen from the
figure, many of the recreationally-valuable islands are located
within this region.

Today, the pollution situation in Boston Harbor remains much
the same as it was in 1968. During 1969, both the Deer Island
and Nut Island treatment plants began year-round chlorination
of liquid effluents at substantially higher dosage rates. This

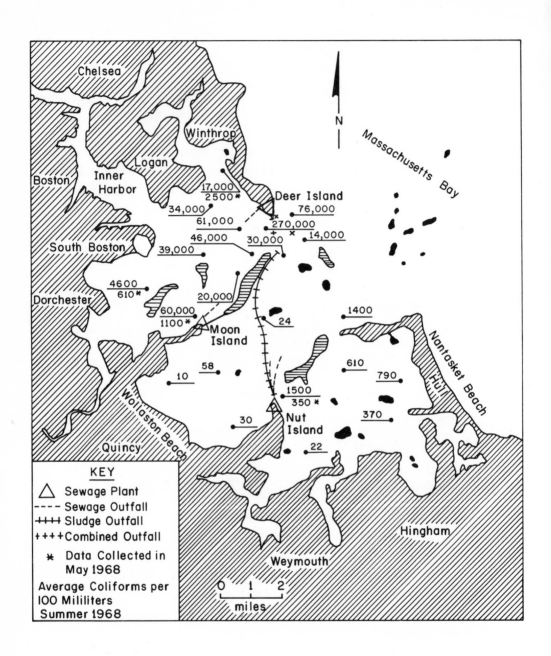

Source: See Reference 12.

Figure 5.4 Bacterial Pollution Densities--Summer, 1968

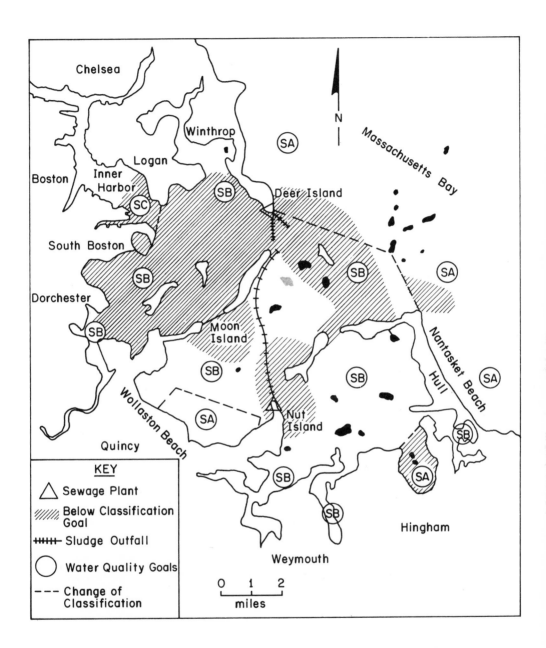

Source: See Reference 12.

Figure 5.5   Harbor Areas Failing to Meet Water Quality
             Goals--Summer, 1968

brought about a noticeable improvement in some portions of Winthrop
Harbor, where six pollution-closed beaches were reopened and
two prohibited shellfish areas were reactivated on a restricted
basis. However, in most other portions of the outer harbor
in the vicinity of the islands, shellfishing is still restricted
and the water continues to be unfit for most recreational activities.
This points out that the most pressing task is yet to be confronted
successfully--that of achieving the water-quality goal of Class
SB for the outer harbor so that the recreational potential of
this region can be fully realized.

2. The <u>Sources</u> <u>of</u> <u>Pollution</u> <u>in</u> <u>Boston</u> <u>Harbor</u>

The major contributing factors which have led to the degradation of the water quality in Boston Harbor are as follows:

1)   Treated municipal sewage
2)   Sludge
3)   Combined sewer overflows
4)   Raw sewage outlets
5)   Oil spillage
6)   Polluted tributary streams
7)   Refuse and debris
8)   Waste from ships and pleasure boats
9)   Federal facility discharges

Treated Municipal Wastes

The single greatest contribution to the pollution in Boston
Harbor comes from the overflow or discharge of raw and partially-
treated sewage from the two major sewage systems operated by the
Metropolitan District Commission (MDC). The harbor serves nearly
two million people as the terminus of their sewer system. A total
of 460 million gallons per day (mgd) of treated sewage is discharged into the Harbor, most of which comes from the MDC's Deer
Island (350 mgd) and Nut Island (110 mgd) treatment plants.[14]
The sludge and sewage outfalls from these facilities are as indicated in Figures 5.3 and 5.4.

Sludge

Sludge consists of semiliquid sewage wastes, grease, oil, tar, sand, gravel, and other diverse solids that are mechanically separated from raw sewage and which decompose slowly on the ocean bottom and tend to accumulate through the years. Sludge contributes to the depletion of dissolved oxygen, can severely disturb bottom fauna, and causes unsightly slicks when carried to the surface. It is known that much of the Boston Harbor bottom is covered by a three-foot-thick layer of sludge.

Combined Sewer Overflows

Raw sewage is often dumped directly into the harbor from malfunctioning or overflowing combined sewer systems which are designed to carry normal dry-weather flows to the treatment plants but which overflow into the harbor during storm conditions. In Boston, there are over 200 relief points, more than 90 of which flow directly to the harbor. Data from a report by Camp, Dresser and McKee[15] shows that nearly fifty percent of the combined storm sewers in the Boston area discharged either most or all of the time in 1967.

Raw Sewage Outlets

Raw sewage is discharged illegally from some shoreline structures, while the Town of Hull continues the practice of dumping untreated municipal wastes into the harbor. The Moon Island facility of the City of Boston has only recently ceased operations of this sort.

Oil Spillage

Occasional oil spillage occurs during oil transfers, especially at terminals on the Chelsea River. Moderate amount of oil sometimes persist in remote sections of the harbor for considerable time periods. The inner harbor is persistently coated with an oily film that affects both the passage of sunlight and oxygen through the air-water interface.

Polluted Tributary Streams

Polluted rivers are a major source of pollution to the
Boston Harbor area. Many industries and cities use rivers as
sewage disposal systems, while malfunctioning combined sewers are
abundant. The most severe contributors are the Charles, Neponset
and Mystic Rivers, with Chelsea and Weymouth Fore Rivers not far
behind. The worst pollution in the entire harbor is at the mouth
of the Charles River in the inner harbor.

Other Contributors

Less major pollution sources include: federal installations,
the Boston Naval Shipyard, the South Boston Naval Annex, the Navy
vessels berthed in the harbor, the Boston Coast Guard Base, the
Nike Ajax site in Hull, watercraft wastes including the dangerous
(because of its long decomposition time and poisonous nature) oil
and tar spills or discharges, and debris and refuse from shoreline demolition.

IV. <u>APPROACHING</u> <u>THE</u> <u>PROBLEM</u> <u>OF</u> <u>CLEANING</u> <u>UP</u> <u>BOSTON</u> <u>HARBOR</u>

Many of the benefits of cleaning up Boston Harbor are hard
to assess in monetary terms, such as increased esthetic enjoyment,
reduced stress on marine ecology, and especially the value of
expanded recreational opportunities for metropolitan residents.
About the only reliable estimate of benefits concerns shellfishing, where, in terms of economic value to the food industry,
the maximum annual loss in Boston Harbor is believed to be approximately 1.3 million dollars.[16] While it may be possible to estimate the monetary equivalence of less boat maintenance or increased swimming and sport fishing (based on the hourly wage or
some similar measure), such estimates are tenuous at best. The
most important sociological value of shoreline recreational resources, now and in the future, belies description in quantita-
tive terms. Consequently, one is faced with difficult tradeoff
decisions regarding the allocation of coastal zone resources
among competing uses.

In the case of Boston Harbor, however, such questions are academic since the decisions concerning the future use of the Boston Harbor islands have already been made within the political process. The City of Boston, the Metropolitan District Commission, the Metropolitan Area Planning Council of the Massachusetts Department of Natural Resources, the New England River Basins Commission, and the Federal Water Quality Administration have all recognized the pressing need to clean up the harbor so that it can be used to its fullest potential as a recreational resource. The Massachusetts legislature has authorized the purchase of all the privately-owned islands in the harbor, to be developed specifically for the purposes of recreation and conservation. The intent is clear and the direction has been established. The task at hand is to find some means of bringing about the stated water-quality goals for the harbor that expends resources in the most economically efficient way. We will address ourselves to a particular segment of this task in the remaining sections.

1. What Has Been Done?

Since the April 30, 1969 conference on pollution of the navigable waters of Boston Harbor,[17] some progress has been made toward achieving the water-quality goals. This progress has recently been reported by the New England River Basins Commission.[18] The most substantive elements of progress to date are as follows:

1) A plan of study to ensure coordination of plan formulation among participating agencies has been developed;

2) The consulting engineering firm of Hydroscience, Inc., has been engaged to develop a mathematical water-quality model and make recommendations for improving the harbor's water quality;

3) Assorted incremental improvements have been carried out including: improved operational efficiency at the Deer Island treatment plant; tidegate repair by the City of Boston; a stormwater detention facility on the Charles River; an oil boom across the Chelsea Creek; debris collection by the Corps of Engineers; and new legislation controlling pollution from watercraft.

These are all certainly steps in the right direction. The
next step is to determine in which areas we should focus our
future efforts to obtain the most productive results.

2. Where Do We Go from Here?

By far the greatest amount of ocean pollution in Boston Harbor results from municipal sewage disposal. For most cities
along ocean fronts the combination of raw, combined overflow, and
incompletely-treated sewage accounts for about 60 to 75 percent[19]
of the overall problem. Even treated wastes contain large
amounts of plant nutrients and harmful bacteria, much of which is
contained in the sludge by-product of the treatment process.
Some recent observations by the New England River Basins Commission of an interim consultants' report on harbor water quality
place primary emphasis on these factors:[20]

> ...The bacterial pollution caused by combined sewer
> overflows and the inflow of tributary streams to the
> Harbor will continue to impair water quality unless
> remedial action is secured.

> ...The treated waste and sludge discharge of the MDC's
> Deer Island and Nut Island waste treatment plants are
> important determining factors in the enhancement of
> water quality levels achieved in the Harbor.

The problem area of combined sewer overflows was studied extensively
in a 1967 report[21] by the consulting engineering firm of Camp,
Dresser, and McKee, which recommended that the least costly
alternative for collecting and disposing of overflows of mixed
sewage and stormwater was a Deep Tunnel Plan. Such a plan would
provide an effective (albeit costly) long-term solution to this
important component of the overall pollution problem. Since
this proposal has been well developed, we have chosen to concentrate our efforts on the second problem area--the effects of
municipal waste treatment facilities--with a focus on the problem
of sludge handling and disposal.

V. MUNICIPAL WASTES AND THE PROBLEM OF SLUDGE

1. Treatment of Municipal Wastes[22]

The overall purpose of waste treatment is to remove or reduce the oxygen-demanding materials, bacteria, plant nutrients,
and suspended solids contained in sewage. Conventional municipal
waste treatment processes are usually broken down into two general categories: primary and secondary. Both processes begin by
dividing the sewage into two components: *sludge* and *liquid effluent* (waste water). This is accomplished through mechanical separation of grease, oil, tar, sand, gravel, and other solid wastes
from sewage through processes such as screening, grinding, scum
removal, and sedimentation. The aggregate of these settleable
solids becomes part of what is called *primary sludge*, a semiliquid
containing 0.5 to 5% solids. With primary treatment, the sludge
undergoes bacterial digestion to reduce organic compounds to more
stable forms, while the liquid effluent is chlorinated to kill
harmful bacteria. Secondary treatment, when used, exposes only
the clarified waste water from the primary process to microorga-
nisms which carry out in a controlled fashion the degradation
process that breaks down organic matter in nature. This process
generates secondary or biological sludge. The combined sludges
must then be treated and disposed of, and how this is done has an
important effect on the quality of receiving waters.

If sludge disposal is carried out properly, the primary and
secondary treatments provide removal efficiencies as shown in
Table 5.3. This table shows that a properly operated primary
treatment facility is capable of removing 35 percent of the oxygen-demanding materials, while the addition of secondary treatment
can increase this to 90 percent. However, *the disposal of digested
sludge through direct discharge into receiving waters greatly
reduces the overall effectiveness of the treatment plant in removing bacteria and oxygen-demanding materials, and considerably
negates whatever nutrient removal there might otherwise be.* The
importance of this factor has been emphasized by the Federal

SEWAGE COMPONENT	REMOVAL EFFICIENCY	
	Primary	Secondary
Biochemical Oxygen-Demanding Materials	35%	90%
Suspended Solids	60%	90%
Nitrogen	20%	50%
Phosphorous	10%	30%
Dissolved Minerals	--	5%
Refractory Organic Materials	20%	60%

Source: L. W. Weinberger, et al., "Solving Our
         Water Problems--Water Renovation and Reuse,"
         in Annals of the New York Academy of Sciences
         136, Art. 5, 131(1966).

Table 5.3   Removal Efficiency of Treatment Processes

Water Pollution Control Administration:[23]

> Sewage treatment in a properly designed and operated
> primary treatment facility is capable of removing 30
> to 35 percent of the oxygen-demanding materials.
> However, unless the nutrients present in waste dis-
> charges are also removed, phytoplankton activity, such
> as that occurring in Boston Harbor, will produce oxygen
> depletions that will continue to endanger the aquatic
> life of the harbor. Adequate secondary treatment of
> sewage can reduce the nutrient content of the waste
> discharge and is capable of removing from 85 to 95
> percent of the organic matter and greatly reducing
> the coliform bacteria. Disposal of the digested sludge
> into the receiving waters increases the amount of nutri-
> ents and oxygen-demanding materials in those waters and
> reduces the overall efficiency of primary or secondary
> treatment facilities.
>
> The Federal Government has not granted funds to the MDC
> for construction of the Deer Island sewage treatment
> facility because of the MDC method used for the dis-
> charge of sludge.

This points to the importance of focusing on the sludge component
of the treatment process, especially in instances where only primary treatment is used, as is the case with both the Deer Island

and Nut Island facilities in Boston Harbor.

2. Sludge

Sludge handling and disposal play a major role in the effective treatment of municipal wastes to minimize water pollution.
This role has been described by the American Chemical Society:[24]

> Handling and disposing of sludges is the single most
> troublesome aspect of waste water treatment today.
> Often it accounts for 25 to 50% of the capital and
> operating costs of a treatment plant. By 1980 the vol-
> ume of sludge requiring treatment will have grown an es-
> timated 60 to 75%, and the increasing costs of labor and
> land that can be used for ultimate disposal will have
> rendered the situation even more difficult.

The primary objectives of sludge treatment are: 1) destruc-
tion of harmful organisms; 2) separation of solids and liquids to
reduce volume; and 3) conversion of organic matter to a relatively
stable form. To accomplish these, five methods[25] are commonly
used:

1)   *Concentration* to initially separate the solids and
liquids, usually through sedimentation or flotation;

2)   *Digestion* by bacteria to decompose organic solids to
more stable forms; also to reduce volume of sludge;

3)   *Dewatering* to reduce the sludge to nonfluid form by
drying on sand beds or vacuum filtration;

4)   *Heat drying or incineration* to again reduce sludge
volume (by removing water) and to sterilize organic solids;

5)   *Final disposal* on land, or in specially-prepared lagoons.

Of these methods, the American Chemical Society has stated that
"anaerobic digestion, followed by dewatering of the digested
sludge on sand beds and disposal as landfill or soil conditioner,
remains a cheap and simple solution to the sludge processing
problem."[26] Some treatment facilities have tried to heat-dry
sludge and sell it as fertilizer or soil conditioner, but the
practice is uneconomical compared to landfill or incineration.

> Heat-drying costs more than incineration, and limited
> demand for the product has made it difficult to get a
> high enough return to offset the increased cost...gene-
> rally the process is considered uneconomical.[27]

The society reports that disposal through combustion has substantial potential in that is seems likely to be able to cope with
all of the sludge disposal problems of the future. In fact, two
out of three of the most common waste treatment processes being
utilized by *new* treatment plants in the U.S. involve some form of
thermal disposal. These most common processes are:[28]

1)   Dewater digested sludge mechanically and use it for
landfill;

2)   Dewater digested sludge mechanically and dispose of it
by thermal means, such as incineration;

3)   Dewater raw sludge mechanically and dispose of it by
thermal means.

All of the sludge-handling processes discussed here reflect
a common aversion to using natural water bodies as receptacles
for sludge. We can understand the reason for this by looking
at some of the relative statistics for digested sludge and effluent
sewage (liquids) at the Deer Island and Nut Island treatment
plants in Boston, as shown in Table 5.4. Note that, *although
the volume of digested sludge is very small compared to the
volume of effluent sewage, coliform bacteria are present in
extremely high concentrations, even higher than the bacterial
content of the total incoming sewage.* This happens because
the digestion tanks provide an ideal environment for the growth
of bacteria. The operation of the digesters is based on the
exposure of organic compounds in sewage to anaerobic bacteria
that reduce these compounds to a more stable form through biochemical decomposition. The temperature is maintained at about 95°F
and is conducive to the growth of both the anaerobic and the
coliform bacteria. Hence, the concentration of coliforms in
the outgoing digested sludge may well be greater than that in
the incoming raw sludge!

CATEGORY	DEER ISLAND	NUT ISLAND
Total liquid effluent flow to harbor (million gallons per yr)	109,123	45,548
Total sludge added to digesters (million gallons/million dry lbs per year)	91.9/62.1	103.6/44.5
Digested sludge withdrawn to harbor (million gallons/million dry lbs per year)	94.8/44.8	$100.0^1/21.5^2$
Approximate coliform concentration of total influent (thousands per 100 milliliters)	100,000-200,000	100,000-200,000
Average coliform concentration of *chlorinated liquid effluent* (thousands per 100 milliliters	∿1.0	∿2.0
Average coliform concentration of *digested sludge* withdrawn to harbor (thousands per 100 milliliters)	(not measured)	$360,000^3$
Maximum coliform concentration of digested sludge withdrawn to harbor (thousands per 100 milliliters)	(not measured)	$700,000^3$

1. Estimate based on approximate 1:1 ratio of digester inflow to digester outflow

2. Estimate based on reported removal efficiency of solids--51.7 percent

3. Based on Federal Water Pollution Control Administration measurements, summer, 1968: see reference 12, p. 52

Source: Metropolitan District Commission Sewerage Division, Fifty-First Annual Report for Fiscal Year Ending June 30, 1970.

Table 5.4   Sewage and Sludge Statistics for Boston Harbor: 1969-1970

Another very significant point to be made about the bacterial content of sewage outfall concerns its expected lifetime. Coliforms and pathogens contained in liquid effluents survive well in fresh water, but their salt-water lifetime is markedly lower. Such is not the case, however, with the bacteria in digested sludge. The digestion tanks are effectively a high salt culture medium due to the leakage of sea water into the main sewage network. Bacteria in this culture can then mutate by a process of natural selection, making them better able to survive in a salt environment, i.e., the ocean. Consequently, unlike their counterparts in the effluent liquids which have a relatively short salt-water lifetime, the bacteria in digested sludge are more sturdy in this regard.

In addition to being a source of bacteria, recall that digested sludge can contribute to the degradation of receiving waters in a number of other ways. These wastes, even after digestion, contain significant amounts of oxygen-demanding materials and plant nutrients. The nutrients contribute to the excessive phytoplankton and other marine growth in many portions of the harbor. Finally, sludge that settles to the bottom of the harbor can have severe long-term effects on the ecology of fauna on the ocean floor. We should note in this regard that no sludge whatsoever is allowable in Class SA and SB waters.

The discussion up to this point has been intended to provide a background of useful information with regard to waste treatment and sludge disposal. We are now prepared to focus attention on the techniques employed by the Deer Island and Nut Island facilities in Boston and the effect that these plants have on the water quality in the harbor.

3. Waste Treatment and Sludge Handling in Boston

We have already noted that the Metropolitan District Commission (MDC) operates two major treatment plants in Boston Harbor, one each at Deer Island and Nut Island respectively. Both of these facilities use *primary* treatment of raw sewage. The treatment processes begin with coarse screening and grit removal,

pre-chlorination, and pre-aeration of the influent sewage.
The waste is then pumped into sedimentation tanks at the main
plants where raw sludge and scum are separated from the liquid
effluent. Prior to the beginning of operations of the Deer
Island plant in 1968, this raw sewage was being pumped directly
into the harbor.

After sedimentation, the *liquid effluent* is subjected to intensive chlorination to destroy harmful bacteria. Chlorine usage
of 10.5 parts per million for about 20 minutes has been extremely
effective in reducing the coliform concentrations of these waste
waters, as the data in Table 5.4 indicate. After chlorination,
the effluent liquid is discharged directly into the harbor at
both Deer and Nut Islands.

Raw sludge, having been separated from the liquid effluent
in the sedimentation tanks, is thickened and then subjected to
anaerobic digestion for approximately three weeks. *There is no
direct chlorination of the digested sludge.* At Deer Island, the
sludge is diverted after digestion back into the main outfall
pipes where it comes into contact with the chlorinated liquid effluent for approximately 10 minutes before reaching the harbor
waters. This serves to kill some of the bacteria present in the
sludge, *but probably not a significant amount since the chlorine
residual in the effluent liquid is (at the point of sludge addition) only about one part per million and the exposure time is
only 10 minutes.* No other disinfection of the sludge takes
place. At Nut Island, the digested sludge is not even exposed to
chlorinated effluent—it is discharged through a separate pipe
approximately four miles out into the harbor. The outfall (with
a coliform density of ~300 million per 100 ml) is just beyond
Long Island, as can be seen in Figure 5.4.

At both Deer Island and Nut Island, sludge is discharged for
approximately four hours a day, and only on the outgoing tides.
However, this by no means assures that the sludge is carried out
to sea, since the mean tidal excursion in Boston Harbor is on the
order of six miles. Hence, the sludge is carried in and out of

the harbor by the tides until it is diluted or settles on the
harbor floor.

It is interesting to note that the combined sludge outfalls
of the Deer Island and Nut Island treatment plants, rich in coliform bacteria nutrient and oxygen-demanding material, form a
"cross fire" of sorts on the central portion of the harbor where
most of the islands are located. As we saw in Figure 5.4, it is
this very portion of the harbor that is well below the water-
quality standards for bacteria that would allow the islands to be
fully used for sorely-needed recreational purposes. This points
to the importance of determining how much of a contribution the
discharge of digested sludge actually makes to the bacterial
pollution in that section of the harbor.

Certainly there are a number of complex factors in addition
to sludge that contribute to the bacterial pollution of the harbor, including combined sewer overflows and contaminated tributary streams such as the Charles River. Determining the relative
magnitudes of these contributing factors is a difficult task requiring the development of a dynamic model to show the influences
of winds, tides, ocean currents, and other factors that determine
the extent of pollution in the harbor. Since such a model was not
available at the time of this investigation, we cannot state with
absolute confidence the role that sludge disposal plays in the
bacterial degradation of Boston Harbor. However, we *can* point
to some data that seem to suggest that sludge does in fact exert
a major influence.

4. The Case Against Sludge

We have previously noted in comparing the coliform counts of
the summers of 1967 and 1968 that the bacteria levels in the harbor showed little or no net change over that period. This is
illustrated in Table 5.5 by data for selected stations in the
northern part of the harbor, as shown in Figure 5.6. While some
stations show an improvement in water quality, others show a
marked degradation, even though the Deer Island treatment plant
had been in operation since May 1968. In its 1969 report the

STATION[1]	COLIFORM DENSITY (#/100 ml)		
	Summer '67[2] (July, Aug.)	Summer '68[2] (J. Jul. Aug. Sep.)	September '68[2]
1968			
BA	47,000	17,000	26,300
BH-26	-	270,000	342,800
BH-27	66,000	76,000	145,900
BH-28	-	14,000	25,300
BH-29	-	6,200	12,000
BH-30	-	13,000	17,700
BH-31	65,000	46,000	43,800
BH-32	-	43,000	71,200
BH-33	-	61,000	117,500
BH-34	-	45,000	80,300
BH-35	-	34,000	61,000
BH-36	31,000	39,000	48,200
BH-37	-	20,000	30,000

1) During the summer of 1968, samples were taken at many more locations than during the summer of 1967. The data shown for 1967 are from stations identical to or in close proximity to the 1968 stations.

2) Data taken during monthly sampling periods is averaged over complete tidal variations, usually for a period of three days.

Source: Same as Table 5.4.

Table 5.5 Average Coliform Densities for Selected Sampling Stations--Summer 1967 and 1968

Federal Water Pollution Control Administration attributed this to the "limited chlorination capacity" of the Deer Island plant, since chlorination during the first two months of the summer of 1968 was intermittent. However, by August 13, water samples taken by the FWPCA indicated that coliform densities *in the liquid effluents* had been reduced to an average of 35,000 per 100 ml. These waste waters comprise 99.9 percent of the total sewage which, prior to the operation of the plant, had been dumped di-

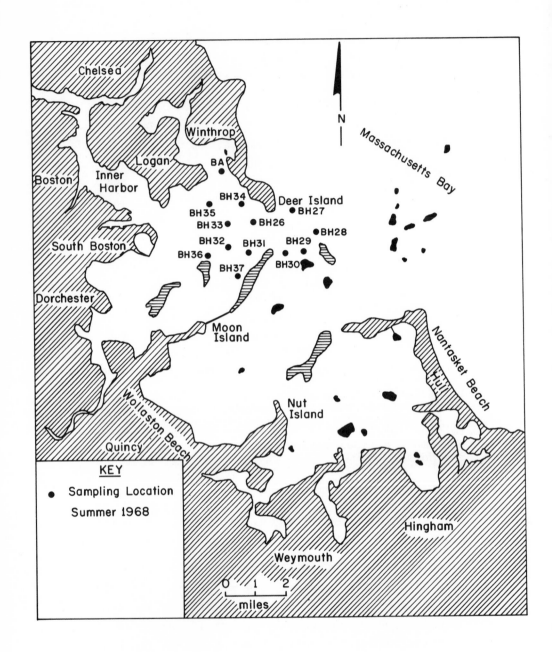

Source: See Reference 12.

Figure 5.6   Selected Sampling Locations in Boston Harbor
Summer, 1968

rectly into the harbor, with a coliform density of about 175
million per 100 ml. It seems strange indeed that even though the
coliform density in 99.9 percent of the sewage had been reduced
by a factor of *five thousand* or so, the bacterial levels in the
waters directly off Deer Island had still not shown any consistent improvement as of late September 1968. It seems illogical
to suspect that further chlorination of the liquid effluent is
what is needed to improve the quality of the receiving waters;
even if the coliforms were brought down to 1000 per 100 ml,
this is still only a factor of 35 reduction as compared to 5000
previously. This leads us to suspect that the problem is located elsewhere, in the other 0.1 percent of the sewage discharge,
i.e., *unchlorinated sludge*. Offhand, this seems consistent with
the possibility that the bacteria in sludge are much better able
to survive in salt water than their counterparts in the liquid
effluent.

Examination of Table 5.5 reveals another very interesting
point. The coliform densities at every station but one in the
northern part of the harbor showed a substantial *increase* between the summer averages and the September averages, *even though
effective chlorination had been achieved in early* August. During
the earlier months of the summer (June, July), only 50 percent of
the incoming raw sludge had been dumped into the harbor as proper
digestion startup was being attempted. However, beginning in
August, *all* sludge was again discharged into the harbor, having
been digested at the treatment plant. Even though the discharge
was on the outgoing tide, the coliform levels measured on September 24, 25, and 26 still showed a marked increase over the summer
average, even though the high September data was included in
those averages. *This again points to sludge as a primary source
of coliforms in the harbor.* To substantiate this, we can go a
step further and examine in more detail the events surrounding
sludge disposal during the summer of 1968.

The FWPCA report of 1969 includes a chronology[29] of sludge
disposal events for the Deer Island treatment plant, beginning

on May 15, 1968 with the commencement of operations of the plant
and running through to the last week of September 1968. The
report also provides[30] coliform counts for a number of sampling
stations in the harbor, taken at various intervals throughout
the summer. In Table 5.6, we relate the chronology of events
at Deer Island to coliform counts at two sampling stations in the
northern portion of the harbor--one in Winthrop Bay (BA) and one
in Dorchester Bay (BB). There seems to be a consistent correlation between the discharge of sludge (whether raw or digested)
and the bacterial levels at the stations noted. Based on these
events we can make the following observations:

1) During the first few weeks of plant startup in May
1968, no sludge was being dumped into the harbor as the digestion
tanks began to fill up. Near the end of May, significant reductions in coliform densities were noted.

2) Beginning on June 1 and running until August 8, varying amounts of raw sludge were discharged into the harbor while
the facility was trying to effect proper sludge digestion. This
appeared to have a *rapid* and *substantial* effect on the degradation of the water quality, as measured on August 8.

3) Beginning on August 9, no raw sludge was disposed to
sea as the No. 3 digester filled up. After only one week or so,
coliform levels had again been drastically reduced.

4) Beginning on September 10, digested sludge was emptied
into the harbor on the outgoing tides. Two weeks later during
the sampling period, bacterial counts were back up to the high
level of August 8.

These results again seem to indicate a strong correlation between
the dumping of sludge and the bacterial pollution of the receiving
waters.

Perhaps we can shed some more light on this situation with a
rough calculation. We have previously noted that the volume of
effluent liquids discharged annually from Deer Island is about
1,000 times the volume of digested sludge dumped into the harbor.

SAMPLING DATE/EVENT	AVERAGE TOTAL COLIFORMS (#/100 ml)	
	Station BA[1]	Station BB[2]
July-August 1967; raw sewage discharged into harbor	47,000	19,000
May 21-23, 1968; no sludge disposal to harbor since May 15, when plant went into operation	2,500	610
July 23-25, July 30-Aug. 1, 1968; 50% raw sludge disposal to harbor since June 1	7,700	1,030
Aug. 6-8, 1968; 100% raw sludge disposal to harbor since Aug. 1	21,300	6,810
Aug. 13-15, 1968; no raw disposal to sea since Aug. 9; chlorinated liquid effluent discharge @ 35,000 coliforms per 100 ml	1,540	(not given)
Sept. 24-26, 1968; digested sludge emptied to harbor on outgoing tide since Sept. 10; continue chlorinated liquid effluent discharge	26,300	(not given)

[1]Station BA is located at the Deer Island flats near
 Buoy "C-3."

[2]Station BB is located in Dorchester Bay between
 Buoys "3" and "N-4."

Source: U.S. Department of the Interior, Federal Water
         Pollution Control Administration, Proceedings--
         Conference in the Matter of Pollution of the
         Navigable Waters of Boston Harbor and Its
         Tributaries, May 1968-April 1969.

Table 5.6 Chronology of Sludge Disposal Events at Deer Island
           and Coliform Counts in Boston Harbor, Summer 1968

Hence, to get an indication of the relative contributions of
these two sources to the total bacteria counts, we should compare
100 ml of sludge to 100,000 ml of effluent liquids. Based on the
1968 data of Table 5.4, there were about 360 million coliforms in
100 ml of digested sludge, while there were about 35 million
coliforms in 100,000 ml of effluent liquids (after chlorination
at 1968 concentrations). Therefore, in terms of yearly output,
the digested sludge contributed approximately *10* times as many
coliforms to the harbor as did the effluent liquid! This ratio
becomes even more pronounced when we consider that increased
chlorine dosages in 1969 decreased the coliform count of the effluent liquids to approximately 1 million per 100,000 ml. Assuming that the increased chlorine residual also reduces the coliform density in the sludge (due to contact with the effluent
liquid before discharge) to about 200 million per 100 ml, then
the contribution ratio becomes *200 to 1*. Even this is very conservative since 1) there is poor mixing and nonuniform contact
between the digested sludge and the chlorine residual in the combined outfall pipes, and 2) we have assumed that 1 ppm of chlorine could kill nearly 50 percent of the total bacteria. *We
should also keep in mind that sludge disposal from Deer Island is
only half the problem--an equivalent amount of sludge is also
discharged in the same general area of the northern harbor by the
Nut Island plant, and this sludge has not even been partially exposed to any chlorinated effluent.* Thus the combined effect
could be that sludge contributes anywhere from 400 to 600 times
the number of coliforms contained in the effluent liquids!

These results lend additional support to the suggestion that
the disposal of digested sludge is a major contributing factor in
the degradation of water quality in Boston Harbor. In the light
of the arguments made in the preceding paragraphs, it would be
interesting to look at one final event regarding the operation of
the Deer Island plant. Beginning on May 1, 1969, increased chlorination of the liquid effluent reduced the average coliform densities from 35,000 to 1,000 per 100 ml. This also increased the

chlorine residual in the liquids to a steady 1 ppm. Subsequent
to this activity, the bacterial levels in Winthrop Bay showed a
marked improvement, allowing the reopening of a number of beaches
and shellfishing areas. While this has been attributed to the
reduction of bacterial density in the effluent liquids, such an
explanation seems unlikely in the light of our previous discussion. Rather, *we feel that the improvement was probably due to
the exposure of the digested sludge--rich in coliform bacteria--to
the increased chlorine residual in the liquid effluent at the
point where the two flows come together.* This seems a reasonable
assertion especially since it is likely that the chlorine concentration has a nonlinear relationship to the amount of bacteria
killed, i.e., the first ppm added provide much more disinfection
than the last. Hence, raising the chlorine residual from practically nothing to 1 ppm could bring about a significant reduction in the bacteria contained in the digested sludge.

None of the arguments that have been made here are conclusive in themselves since we have had access to a limited amount
of data and other informational resources. Taken together, however, the indications that sludge is a major contributing factor to the bacterial degradation of Boston Harbor are much too
strong to be ignored. Certainly a much more detailed analysis
will be required before the true nature of this problem can be
fully understood. Such analysis must determine the biological
characteristics of the bacteria in digested sludge, the relationship between chlorine usage and effluent disinfection, the dynamic behavior of these bacteria in the salt-water harbor, and
the contribution that sludge makes relative to other sources
such as sewer overflows and polluted tributaries. *If the suggestions we have made are proven correct, then it is of primary
importance that the dumping of digested sludge into the harbor
must be discontinued if the water-quality goals are to be met
within the foreseeable future!*

In the next section, we will evaluate in a preliminary way
some possible alternatives to harbor dumping.

VI. SOME POSSIBLE SOLUTIONS

As a first step in finding a suitable disposal scheme for
digested sludge which avoids dumping in the harbor, we have
looked at preliminary cost estimates for four alternative proposals:

1)   Pipe undried sludge to some landfull area (~20 miles)

2)   Barge the undried sludge farther out to sea (~15 miles)

3)   Pipe undried sludge farther out to sea (~10 miles)

4)   Dry and store sludge at Deer Island

The first plan would transfer Nut Island sludge to Deer
Island and then the combined undried sludge would be piped to
some landfill area 20 miles away. The primary difficulty with
this plan is finding a suitable landfill site (at a feasible
cost) in a tight metropolitan land market. Also, the undried
sludge would have to be processed at the site to avoid offensive
odors. The second plan is slightly less costly than the first,
but we feel that careful study is needed with regard to the uncertainties of the effects of sludge on the ecosystems of the ocean.
The third plan suffers under the same uncertainties as the second
in addition to being extremely costly. On the basis of our estimates then, it appears at present that the fourth proposal provides the best choice among the alternatives considered. Certainly there are numerous other possible alternatives to sludge
disposal through harbor dumping that may be attractive in the
near future and which merit careful attention. One promising
technique that we have noted is disposal through thermal means
such as incineration. But at present the most widely-used method is drying and storage on land:

> The most economical method of sludge disposal depends
> on local conditions...and methods other than combustion
> seem likely to retain their utility for a long time.
> The ability to evaluate alternative disposal methods
> soundly will require thorough investigation of such
> questions as the value of liquid, dried, or composited
> sludge as a fertilizer or soil conditioner; underground
> disposal as in abandoned mines; and pipeline transporta-
> tion of sludges.[31]

Under the Deer Island landfill scheme,[32] the sludge would be dried
on sand beds or by using mechanical means and then deposited on a
landfill site at Deer Island. An area of 300' by 200' by 5'
would last approximately 10 years or so before it would become
necessary to truck some of the accumulated sludge away. A wall
could be built around the site to retain the sludge, while a roof
may be needed to keep out the rain. A rough estimate of the
costs is as follows:

```
Drying equipment...$3,000,000
Site development... 4,000,000
Nut Is. to Deer
  Is. Connection.... 1,000,000
TOTAL CAPITAL COST.$8,000,000

Operation and Main-
  tenance..........$  110,000
  (includes periodi-
  cally trucking
  away accumulated
  sludge @ 10¢ per
  ton-mile)
```

These cost estimates have been found to be in basic agreement
with those found in Reference 32 which describes a similar sludge
treatment scheme.

VII. SUMMARY AND CONCLUDING REMARKS

Boston Harbor is a uniquely valuable natural asset to the
people of the Boston metropolitan area and of New England. This
value lies in its intrinsic suitability for recreation and other
noncompeting water-related uses. However, unless the serious
problem of water pollution in the harbor is overcome, the full
potential of the area can never be realized. This fact has led
to the establishment of water-quality goals and a management plan
for the abatement of pollution in the harbor. This chapter is
intended to serve as an input to this planning and management
process.

The most important sources of pollution in the harbor are
municipal sewage and sludge from the treatment plants, and raw
sewage from combined sewer overflows and illegal dumping. Al-

though the continuous operation of the Nut Island and Deer Island
treatment plants have had some favorable effect on harbor water
quality in certain areas, the practice of direct discharge of
digested sludge into the harbor has substantially reduced their
effectiveness. *There is substantial evidence to suggest that
this sludge has an extremely degrading and widespread effect
on the bacterial quality of the water.* First, sludge dumped
into the northern portion of the harbor from Nut Island and Deer
Island contains approximately 500 times the amount of coliform
bacteria present in chlorinated liquid effluents, even though the
volume of these effluents is nearly 1,000 times that of the
sludge. Second, a chronology of sludge disposal events during
the summer of 1968 seems to indicate a strong correlation between
the disposal of sludge (raw or digested) and the coliform densities in the northern sector. Third, the fact that increased
chlorination of the liquid effluents in 1969 resulted in substantial reductions of coliform densities in some parts of Winthrop
Bay suggests than an even greater improvement might be effected
if sludge were not dumped into the harbor, since its effect on
the coliform population is so much greater than that of the
chlorinated liquids.

All of this evidence is consistent with the fact that proper
sludge handling and disposal is widely recognized as an important
component of any effective pollution-abatement program. *We feel
that the current method of disposing sludge in Boston Harbor is
one important reason that the water is still of unacceptable
quality, especially near the islands which unfortunately are
located in the vicinity of the sludge outfalls.* Thus the question of sludge disposal must be one of top priority in the management plan for achieving the water-quality goals for the harbor
in the near future.

We have examined in a preliminary way some alternatives to
disposing sludge in the harbor, including drying and storage at
Deer Island. This analysis was not intended to be complete; we
realize that there may be other more efficient ways to handle

sludge, perhaps through incineration or even direct chlorination. Our primary purpose has been to focus attention on the immediate need to attach a high priority to the entire question of sludge and its effect on the harbor water quality.

In looking at the entire sludge situation, the first and most important task is to understand to what extent it contributes to the bacterial pollution in each sector of the harbor. We need this information to evaluate the relative costs and benefits of various long-term cleanup alternatives. One such alternative, recommended in the Camp, Dresser, and McKee study in 1967 as the least costly long-range solution, is a Deep Tunnel Plan that would dump all incoming sewage 9-1/2 miles out into the ocean from Deer Island. The only treatment the sewage would receive would be heavy chlorination. This tunnel would accommodate the normal sewage flow as well as all storm overflows. While the plan might drastically reduce pollution in the harbor, the cost is great: approximately 1 billion dollars when capitalized for 30 years at 5 percent. On the other hand, consider the capitalized costs for the sludge storage plan: approximately 42 million dollars. If in the vicinity of the islands sludge is the source of 50 to 75 percent of the bacterial pollution, *perhaps the water-quality goals could be met in that region by concentrating efforts on finding a better sludge disposal scheme.* Or, a combined program of tidegate repair and sludge disposal might achieve a level of pollution abatement that would allow the harbor to be reopened on a broad scale for recreational and shellfishing purposes. *In other words, compared to the costs, the payoff may be very large if an effective sludge disposal scheme were to be implemented. Thus we strongly urge that the issue of sludge disposal be given careful consideration in the ongoing efforts aimed at cleaning up Boston Harbor.*

REFERENCES

1. U.S. Department of the Interior, Federal Water Pollution
   Control Administration, Report on Pollution of the Navigable Waters of Boston Harbor (1968).

2. The Sierra Club, Eastern New England Group, "Facts about
   Boston Harbor," No. 4, December 1969.

3. Commonwealth of Massachusetts General Law, Chapter 742,
   H4884.

4. New England River Basins Commission, Boston Harbor Coordinating Group, Progress Toward Achieving the Water Quality
   Goals for Boston Harbor (1970).

5. See Reference 1, p. 47.

6. Ibid., p. 50.

7. Ibid., p. 51.

8. Ibid., p. 4

9. Ibid., p. 53.

10. Ibid., p. 55.

11. The Boston Sunday Herald Traveler, December 17, 1967,
    p. II-9.

12. U.S. Department of the Interior, Federal Water Pollution
    Control Administration, Proceedings--Conference in the
    Matter of Pollution of the Navigable Waters of Boston Harbor
    and its Tributaries, 2nd session, April 30, 1969.

13. Ibid., p. 49.

14. See Reference 1, pp. 18-20.

15. Camp, Dresser & McKee, Inc., Report on Improvements to the
    Boston Main Drainage System, Vols. I and II, HUD Project
    No. P-3306, September 1967.

16. See Reference 1, p. 2.

17. See Reference 12.

18. See Reference 4.

19. See Reference 15.

20. See Reference 4, p. 9.

21. See Reference 15.

22. The American Chemical Society, Cleaning Our Environment:
     The Chemical Basis for Action, Washington, D.C. (1969).

23. See Reference 1, p. 20.

24. See Reference 22, p. 111.

25. Ibid., pp. 112-118.

26. Ibid., p. 115.

27. Ibid., p. 118.

28. Ibid., p. 119.

29. See Reference 12, p. 24.

30. See Reference 12, Appendix B.

31. See Reference 22, p. 119.

32. This scheme is similar to the Porteous process as described
     in Robert Sherwood and Frank P. Sebastian, "A New Method
     for Sludge Treatment in Industrial or Municipal Wastewater
     Treatment Facilities," the BSP Corporation (Great Britain).

CHAPTER 6

REGIONAL GOVERNMENT IN NEW ENGLAND:
A PROTOTYPE

by

Robert Field, Jr.
Sandra Lynch
Richard Morse, Jr.
Robert Wolfe

ABSTRACT

All of the environmental problems discussed in the accompany-
ing articles are aggravated by the frequent inability of the
political system to mobilize effective efforts to combat them.
This happens because regional decisions are generally formulated
by non-regional processes, leading to inefficiency, poor planning,
and limited solutions for problems too important to be approached
on a piecemeal basis. In approaching this morass, we have used
three principles as guidelines: (1) Government should be administered on a plane high enough to accommodate coordinated lower-
level implementation of policies; (2) total family income should
be recognized as the only criterion for effective and equitable
taxation; and (3) efficient land allocation should be fostered
by more conscious application of existing control mechanisms.

Consideration of these issues has led us to advocate the
following: (1) Elimination of the multiplicity of governments;
(2) creation of a regional government to administer interstate
problem solving; and (3) overhauling the existing property tax
to emphasize land-use management and other social goals, such as
pollution abatement. We suggest that this can be accomplished in
part through a three-phased program to establish a regional government and through implementation of a new property tax.

CHAPTER 6

REGIONAL GOVERNMENT IN NEW ENGLAND:

A PROTOTYPE

I. INTRODUCTION

As attention focused on the problems of the cities during
the last decade, one federal committee recommended that urban
problems be given regional consideration:

> The most pressing problem of local government in metro-
> politan areas may be stated quite simply. The bewil-
> dering multiplicity of small, piecemeal, duplicative,
> overlapping local jurisdictions cannot cope with the
> staggering difficulties encountered in managing modern
> urban affairs. The fiscal effects of duplicative sub-
> urban separatism create great difficulty in provision
> of costly central city services benefiting the whole
> urbanized area. If local governments are to function
> effectively in metropolitan areas, they must have suffi-
> cient size and authority to plan, administer and provide
> significant financial support for solutions to area-
> wide problems. (Committee for Economic Development,
> Modernizing Local Government 44, 1966.)

The existence of 146 government entities within the metropolitan Boston area illustrates that New England particularly suffers from a maze of fragmented jurisdictions. The second most
urbanized region[1] in the United States is New England where,
in 1960, 76.4% of the total population was urban.[2] Because of
the compactness of the New England region, the resulting sprawl
leapfrogs state boundaries, creating interstate problems of
increasing complexity.

The resulting economic dislocation distorts the difficult
problem of finding sufficient revenues to finance public projects. As the Advisory Commission on Intergovernmental Relations
(ACIR) noted:

> The multiplicity of governments and its political corol-
> lary "home rule" can work against the most efficient
> allocation of resources--a "surplus" situation in one
> community ordinarily will finance projects of increasingly
> lower priority rather than underwrite a high priority
> function in a neighboring community confronted with a

"deficit" situation. It should also be noted that the
multiplicity of local governments creates a political
milieu that makes state equalization efforts more costly
than efficient. In order to help the poorer districts
or communities, it is usually necessary to provide a
measure of aid to all districts including the most
wealthy.[3]   (Emphasis added)

This fragmentation of tax bases generates dangerous fiscal disparities.[4] At the same time that central cities are faced with
the continuing need to spend large amounts per capita for public services, their tax bases are being eroded. New England re-
ceives over one-half of its revenue from the property tax, which
of all taxes is the most inequitable and least responsive to
economic growth. The finances of the New England states reveal
some consistent patterns:

(1)   Receipts from the Federal Government to be applied
against the four major functional areas--education, highways,
public welfare, and health and hospitals--were substantially
below the national average.

(2)   State-level payments for education (26.5%) were significantly below the national average (40.7%) with the result that
the local share of education costs among the states (67.6%) far
exceeded the national average (52.0%).

(3)   In all six states, per $1,000 of personal income, edu-
cation was by far the largest expenditure item in 1967-68.

(4)   The state's share of expenditures for public welfare
and health and hospitals greatly exceeded the national average.

(5)   The net result of the state-local relationships described in items (2)-(4) above was that state aid in New England,
as a percentage of local general revenue in 1966, was only 21.2%,
compared with a national average of nearly 31%.

(6)   Thus the traditional New England emphasis on strong
local government resulted in a reliance on local property taxes
for 53.4% of all state-local revenue from taxes, compared with
a national average of only 43.5%.[5]

(7) Conversely, non-property taxes comprised only 1% of
total local taxes, compared with a national average nearly 13
times as large.[6]

Criticisms of the property tax are legion. Disadvantages
of the tax include the fact that, in order to attract taxpaying
users, localities are forced to develop land for residential and
commercial use which might otherwise be retained for recreational
purposes. In addition, the property tax is difficult to administer. As part of our study, we shall examine a revision of the
tax structure with particular emphasis on revising the property
tax.

The Massachusetts Constitution[7] as well as statutory[8] and
case law make it clear that property is to be assessed and taxed
at full "fair cash valuation."

> This means fair market value, which is the price an
> owner willing...to sell ought to receive from one wil-
> ling...to buy.[9]

It has been held to be wholly illegal to assess land at less
than full valuation.[10] Nevertheless, the statewide ratio of
assessed value to sales price has been determined to be only
36.7%[11] Perhaps the greatest disadvantage of the property tax
is that a large percentage of land may be tax-exempt. It has
been estimated that 60% of Boston's real property is tax-exempt
since many schools, hospitals, and religious organizations are
located within the city.[12]

Boston will feel financial strains in the future if the
property tax continues to be a major source of revenue. This
prognosis is supported by two statistics. Of the 37 largest
Standard Metropolitan Statistical Areas (SMSA's), Boston is
the fifth smallest in land area.[13] At the same time, per capita
total general expenditure by the central city is the fourth
largest of the 37 areas.[14] Such statistics have led one group
to conclude that:

> ...economic growth of the City of Boston will be severely
> constrained, even with the changing economic structure of

its economy and the upgrading of jobs and income that may
be expected to accompany the growth of business, personal
and government service activities, as manufacturing and
trade continue to move out....[P]ublic expenditure needs
of the City of Boston, already overwhelmed by outlays for
health, welfare and safety, allowing less than adequate
margins for education, transportation and housing, are
soaring as the City of Boston continues to provide for
the bulk of the area's poor, needy, and disadvantaged,
and as standards of public service for social welfare
rise. To fulfill the potential for the liability of
the City of Boston economy, in these circumstances,
[will] require measures at the state and national level
to redress, in part, the fiscal and economic disparities
presently confronting the City of Boston.[15]

The importance of Boston's problems for New England is seen when
one recognizes that Boston serves at least three states as a
major financial and cultural center. The fiscal and governmental problems of the city should be viewed in an interstate
context.

At best, the tax complex of the interstate metropolitan area
"is difficult, if not impossible, to analyze. As long as there
is a substantial interchange of people between the parts of an
area belonging to the different states, it is unlikely that there
will be any close correlation between the payment of taxes and
the reception of benefits from public expenditures."[16] As a
consequence of this inequitable distribution of the revenue dollar, wrong-way migrational patterns of business and people have
been accentuated, thereby forging "a white, middle- and high-
income noose around the increasingly black and poor inner city;
and [subjecting] much of rural America to a continuing course
of gradual erosion."[17]

The problems of financing government are intertwined with
the settlement and land-use patterns of New England. *It is a
major premise of this paper that planning mechanisms must be
institutionalized in government to deal with the problems of inefficient land use and erosion of tax bases.* Consideration must
be given to the social imbalance resulting from the black-core
city/white-suburb phenomenon. In addition, action must be taken
to restore the ecological balance of the land and waters of New

England, long ago disturbed in the process of industrialization.
New England's open spaces, one of its prime attractions, must
not be usurped from displacement by unplanned and poorly-con-
trolled development.

Existing mechanisms to control land use are inefficient.
The prime mechanism, zoning, instead of following and implementing planning, often precedes and provides the structure within
which planning is done.[18] Although the New England states indivi-
dually have authorized the establishment of planning agencies,[19]
it is felt that the problems which the planners must face do
not end at the boundaries between states. More centralized
planning is needed. Land-use patterns will also be influenced
by federal transportation, labor, power, and foreign trade regulations. The logic of desirable development demands that the
governmental and planning functions for the New England region
be centralized.

Centralization of power refers to "the transfer of effective power of political decisions to higher governmental levels
encompassing wider geographic area."[20] Yet, "at bottom 'centralization' is no more than an attempt rationally to relate
governmental forms and institutions to the geographic breadth
of the public need for uniform regulations and minimum service
standards...A centralized government can be a responsible and
responsive government, and a decentralized government...can
yield irresponsibility as well as diffusion of poerwer."[21]

Regional decisions are now being formulated by nonregional
processes. This has led to inefficiency, poor planning, and
limited solutions for problems too important to be approached on
a piecemeal basis. In approaching this morass, three principles
will be used as guidelines:

(1) Government should be administered on a plane high
enough to accommodate coordinated lower-level implementation;

(2) Total family income should be recognized as the only
criterion for effective and equitable taxation;

(3) *Efficient land allocation should be fostered by more conscious application of existing control mechanisms.*

The basic areas of concern, as already identified, are *government structure, land use,* and *tax structure.*

Consideration of these areas of concern leads us to advocate the accomplishment of the following:

(1) Elimination of the multiplicity of governments;

(2) Creation of regional government to administer inter-state problem solving;

(3) Overhauling the existing property tax to emphasize land-use management and other social goals, such as population control.

This may be accomplished in part through a phased program to establish a regional government and through implementation of a new property tax which would operate more efficiently to allocate land for use. Governmental reform promoting fiscal responsibility would come in three stages.

The first stage, to be completed by 1980, will include intrastate governmental consolidation; establishment of a regional compact; imposition of a city payroll tax; and full-value property tax assessment. Looking to political feasibility, only low-order intergovernmental cooperation is proposed. In order to foster a climate favorable to increasingly non-local control in later stages, several more farreaching mechanisms will be initiated in the first stage.

Once these proposals have been implemented, stage two will commence and will include reform projects establishing Metro- and Sub-Regional governments; modification of intrastate institutions; and refining of revenue measures. Stage two should be completed by the year 2000.

The third stage of reform represents the culmination of previous efforts to consolidate government and should be completed in the years 2015-2020. This stage envisions the comple-

tion of the Sub-Regional and Metropolitan governments which cross
state lines. State and federal constitutional amendments are
advocated, authorizing states to consolidate their powers into
regional governments.

II. STAGE ONE

1. Intrastate Governmental Cooperation

The minimum level of cooperation urged is the voluntary
association of existing governments into localized Councils of
Governments.[22] There should be little opposition to this proposal because it does not threaten existing governmental units
and because it requires neither enabling legislation nor popular
referendum. Such COG's would provide a forum for the exchange of
views on common problems and should lead to jointly-sponsored
legislation, coordinated planning, and cooperation in some
governmental activities. Because such councils would be voluntary and their decisions not binding, they are recommended for
use primarily in rural areas where there is now no inter-local
cooperation.

Inter-local cooperation may also be accomplished without
altering existing governments through contractual relationships.[23] An example is contracting between cities for reciprocal
municipal services such as fire protection and water supply.
Such cooperative contracting is most appropriate for any commodity service or proprietary function which (1) lacks need of
major substantive discretion; (2) involves standardized and accepted performance methods; (3) requires specialized professional
or technical qualifications; and (4) has a comparatively stable
demand.[24] Statutes authorizing such contracting are in existence.[25]

City-county consolidation is not recommended. The county
is an anachronistic governmental unit which should be abolished.
A higher level of cooperation can be achieved through the formulation of *special districts*. Special districts cut across territorial
lines but do not replace governmental units. They assume cer-

tain government functions as public corporations or as quasi-
municipal corporations.[26] These special districts should be
distinguished from authorities such as the Massachusetts Bay
Transit Authority (MBTA). While the accomplishments of the MBTA
and the Metropolitan District Commission are recognized, it
is felt that such independent authorities are nonresponsive
to public will, hampered by the lack of general taxing powers
and, inevitably, jealous guardians of their own delegated powers.
A proliferation of such authorities could lead to overlapping
and working at cross-purposes.

For heavily-urbanized areas such as the Boston and the
Providence Metropolitan Statistical Areas, *federations* of governments are urged. The governing councils of such federations
could be elected from the constituent cities and towns. Council
decisions would bind federation members. An executive, prefer-
ably from an outside area, would be elected by the council.
The Home Rule Amendment to the Massachusetts Constitution authorizes the legislature to create such metropolitan or regional
entities and to grant them powers to tax and borrow.[27] The
metropolitan federation would fulfill regional functions such
as property assessment, debt borrowing, libraries, sewage and
garbage disposal, pollution control, housing for the elderly
and poor, etc. Local functions would be retained by city and
town governments. The functions of the independent authorities
would be assumed by the metropolitan federation.[28]

2. Interstate Regional Cooperation

The use of two currently available tools--the Interstate
Compact and the Urban Development Corporation--to tackle interstate problems is proposed. The problems of metropolitan man
frequently straddle provincial or "state" lines, but still remain
subnational in scope.

Interstate Compacts

Although use of the compact device has produced few outstanding examples of success, the compact could emerge into a

powerful and expedient governmental tool during the next few
decades. To be successful the compact should have strong financial powers and a well-developed political accountability and
responsiveness.[29] The use of interstate compacts is sanctioned
by the United States Constitution, which authorizes a State to
"enter into any Agreement of Compact with any other State" with
"the Consent of Congress."[30]

The interstate compacts currently employed with varying
success may be classified as follows:

(1) Natural resource development (or public welfare)
     compacts:

     (a) Subject matter of enduring concern to the
          whole state;
     (b) User charges are insignificant;
     (c) Reasonable and politically acceptable to
          resort to general state revenues for costs
          above and beyond those born by the Federal
          government; and
     (d) Informal federal involvement.

     Example: *Atlantic and Gulf States Marine Fisheries Compact*

(2) Regulatory compacts:

     (a) Local focus and small costs;
     (b) Governmental in nature;
     (c) General budget of signatory states carries
          the cost burden.

     Example: *Washington Metropolitan Area Transit Regulation
               Compact*

(3) Self-sustaining proprietary local service compacts:

     (a) Financial burden carried by revenue bonds and
          user charges;
     (b) Major objective is public service.

     Example: *The Port of New York Authority*

(4) Non-self-sustaining proprietary local service compacts:

     (a) Designed for large-scale projects;
     (b) Revenue bonds and user charges not expected to
          carry all the burden.[31]

     Example: *Delaware River Basin Compact*

The compacts which have been entered into by the New England
states concern planning, radiological health, water pollution control, and interstate cooperation commissions.

*The non-self-sustaining proprietary local service compact
is proposed for use as an intermediate device for integrating
the New England region.* The efforts of the compact would be
directed to nonprofit, socially-oriented objectives. Such a
New England States Compact would be administered by a full-time
Board of Regional Directors, two from each state, and a permanent administrative staff. During Stage One, the Board would
be responsible for promulgating Regional Development Plans for
New England and engaging in specific study projects. It would
also coordinate plans proposed by the Regional Development Corporations, to be described later.

The problem of massing public support for the Board's plans
and the necessity of insuring the Board's responsiveness dictate
that the Board be cognizant of the following when formulating
policy:

(1) Tax Level Differences. Among those likely to oppose
change are the people who now benefit from the differences in tax levels that characterize almost all
metropolitan areas....To eliminate or narrow the
tax differences by governmental restructuring would,
of course, benefit some metropolitan residents; but
it would increase the taxes of others--and the latter
group is likely to include influential members of the
area's power structure.

(2) Social Disparities. Poor and disadvantaged people,
including a considerable proportion of Negroes and
other ethnic minorities, tend to be concentrated in
"poverty areas." The central cities typically have
far larger proportions of such "high-cost citizens"...
(therefore the well-advantaged) groups may well fear
that governmental restructuring--whatever its possible advantages in other ways--will considerably
reduce their political muscle.

(3) Established Interests. Ongoing governmental arrangements accumulate a host of persons who rely heavily
on the continuation of the status quo....For those
people the prospect of major structural change at

best involves uncertainty, and at worst the possible
loss of familiar advantages of status or economic
benefit.

(4) <u>Public</u> <u>Uncertainty</u>. Most metropolitan residents
lack close acquaintance with the local governments
that serve and tax them...their concern is not likely
to promote structural change unless they can be
convinced that:

    (a) Existing organizational arrangements con-
        tribute seriously to the problems involved;

    (b) Other kinds of action--such as more grants
        from the state or federal government--
        would be inadequate; and

    (c) The proposed structural change offers prom-
        ise of major improvement and is clearly better
        than any available alternative.[32]

It is suggested that the creation of a powerful interstate
compact will have a twofold effect. First, it will enable our
existing governmental structure to meet the environmental and
social challenges with a degree of positivism and responsibility that will permit massive changes during the next two decades. All New England must contribute to the solution of the
problem which is so easily identified and dismissed as being
unique to Massachusetts or Rhode Island, i.e., "metropolitanism."
Second, a strong regional compact which can be shown empirically
to operate efficiently will serve as a catalyst to the future
consolidation of traditional state services and functions into
a regional type of government--subnational in nature, yet fiscally strong enough to fund regional projects, thereby protecting regional and local interests in a "new federalism."

The compact would be funded from direct contributions from
federal grant-in-aid programs[33] and from state support in satisfaction of the contractual compact. The compact would not
represent a new federal subdivision; its authority would be pre-
dicated solely upon that of the founding states.

Since compacts are the products of coordination of interstate interests and can be drafted with infinite variation, they

provide a very practical first step in reflecting government
response to social and environmental pressures. The compact
appears to be an excellent vehicle for governmental officials
to promote positive social change without engaging in political
suicide.

Regional Development Corporations (RDC's)

    While overall planning is to be done by the Board of Region-
al Directors, specific problems will be dealt with by a
Regional Development Corporation. All of the New England
states have adopted legislation creating Development Credit
Agencies and Industrial Bond Plans. This legislation parallels
a national trend:[34]

> Private development credit corporations operate under
> the following general scheme. After incorporation the
> organization issues its stock..., and when a stated
> amount of capital has been paid in, the corporation is
> permitted to start business. Because the corporation
> is designated to provide a source of credit not else-
> where available, and not to compete with existing credit
> sources, prospective borrowers may have to show proof
> that they have been refused loans from commercial sources.
> Lending funds are provided to the corporation at a low
> interest rate by its non-stockholding members, which are
> traditional commercial credit agencies that have agreed
> to make funds available on call in return for obliga-
> tions of the corporation. The limit of each member's
> lending capacity is set at a small percentage of its
> total capital and surplus, and all calls upon the members
> are required to be prorated in relation to the loan limit
> of each. Loans by the corporation are at an interest rate
> slightly higher than that paid to the members, and are
> secured by mortgages on the property or by the stock of
> the borrower. Total obligations of the credit corpora-
> tion are generally limited to a stated multiple of its
> paid-in capital surplus, and loans to any one borrower are
> similarly limited to a percentage thereof.[35]

In Massachusetts, specifically, the agency[36] authorized to
issue bonds is obliged to "promote, stimulate, and advance the
business prosperity of the Commonwealth...to encourage and assist through loans, investments, or other business transactions,
in the location of new business and industry...."[37] The constitutions of the New England states require that such expendi-

tures be for a public purpose.[38] This requirement has been re-
laxed in recent years.[39]

There is no apparent federal constitutional impediment to
the creation, through compact, of interstate Regional Development Corporations vested with authority to act as credit agencies and to issue industrial bonds. They could be utilized to
implement specific objectives of the compact--for instance, low-
cost housing construction and mass transit operation, as well
as real estate market control for housing and industry.

The fiscal operations of the RDC would be run in generally
the same manner as in existing state credit agencies. Although
such credit corporations have had varied success on the state
level, we submit they could be effective on the regional level
if properly administered.

Directed in such a way as to assure coordination with an
overall Regional Plan, each RDC would itself adopt a master
plan. Through vote of both legislative houses, individual
states would have veto power. The RDC's would thus function
within the compact as an operating body to implement the Regional Plan through financing, land acquisition, and technical assistance to private industry.

       3. Local Financing: City Payroll Tax and Total Income
          Assessment

As was stated earlier, over one-half of all local and state
revenue in New England is provided by the property tax. Yet
such a tax is inelastic,[40] particularly when compared with the
automatic growth characteristics of the progressive income tax.
Heretofore, the property tax has been able to keep pace with
revenue requirements through increased assessments and escalation of rates, assisted by an unprecedented expansion in construction.[41] However, it now appears that these tax rates have
approached the limit where they constitute *confiscation*, while
environmental considerations may limit the rapid growth of new
construction.

That the property tax is inappropriate for financing cities
such as Boston was demonstrated earlier. The property tax has
been maintained in its historic form because it is a stable
source of revenue. In a rural society, the value of a family
residence served as a fairly good proxy of the ability to pay
taxes. But in a modern urban society, total household income
is a more precise measure of taxable capacity.[42]

In Stage One, it is proposed that a broader tax base be
achieved through imposition of a payroll tax on all wage earners
within the core cities of Boston and Providence. All wages
above federal poverty level would be subject to taxation at a
low progressive rate.[43] Although this is not consistent with
the desirable objective of taxing total income, the payroll tax
is advocated as a method of providing the core city with much-
needed revenue.

The efficacy of municipal income taxes (a step beyond the
payroll tax) has been demonstrated--at least 17 cities of over
150,000 in population have enacted such taxes.[44] In several
cases the rates are different for residents and nonresidents,
while some cities (New York) have progressive rates. For the
larger central cities the tax offers an equitable, productive,
and administratively feasible source of revenue.[45] It also
allows the city to derive some revenue from nonresidents who
earn income in the city and use city services. A constitutional
amendment would be required in Massachusetts to implement such
an income tax.[46]

To recapitulate, Stage One is to be completed by 1980 and
proposes:

   (1) Instrastate governmental cooperation;
   (2) Establishment of a regional compact;
   (3) City payroll tax;
   (4) Total income assessment.

III. STAGE TWO

1. Governmental Consolidation

If local governments are to function effectively in metropolitan areas, they must be of sufficient size and authority
to plan, administer, and provide financial support for solutions
to areawide problems. To overcome decentralization and provide
sufficient power to deal with metropolitan problems, the Boston
and Providence federations should now be formalized into *Metropolitan Governments*. This would effectively eliminate the
cities and towns within the metropolitan area as decision-making
and revenue-allocating entities. Governing council members
would now be elected from districts of equal population.[47] The
independent authorities, the Massachusetts District Commission,
the Massachusetts Bay Transit Authority, and the Massachusetts
Port Authority all would be subsumed as departments of the Metropolitan Government of Boston. Geographically, the Boston Metropolitan Government should encompass at least the 78 cities and
towns which now compose the M.B.T.A. territory.

Most activities now undertaken by county governments should
be assumed by the metropolitan governments and by the less urban
councils of governments. Some county functions such as penal
institutions, courts, and agricultural schools should be assumed by the New England Compact.

The councils of the metropolitan governments of Boston and
of Providence should be assisted by advisory committees of experts in aspects of urban affairs. Each advisory committee
would be composed of 15 members, one-third each selected by the
governor, the council executive, and the constituent electoral
districts.

Less urban areas would now be governed by institutionalized,
mandatory councils of governments. The New England states would
then in effect be divided into *Sub-Regional Areas* (SRA's), governed
either by metropolitan governments or councils of governments.
Each SRA government would possess regional planning, land use,

and tax assessment powers. The boundaries of the SRA's would
be determined by the state legislatures with the consultation of
the Board of Regional Directors of the New England Compact.

The SRA's should be established in accordance with the following criteria:

(1)   The natural geography;

(2)   Existing land use
      (a)   Industrial
      (b)   Population concentrations
      (c)   Recreational and natural resources
      (d)   Urban/rural interface;

(3)   Current trends in legislation concerning intrastate
      governmental consolidation efforts;

(4)   Existing interstate cooperation and consolidation
      efforts;

(5)   Existing federal land-use policies and restrictions;

(6)   Socially- and ecologically- desirable conditions
      (a)   Balanced urbanization
      (b)   Tolerable pollution levels
      (c)   Efficient regional transportation
      (d)   Accessibility of recreational and natural
            resources.

In order for the SRA's to respond to regional concerns,
it may be necessary to establish subregional governments which
cross state lines--for instance, in the Springfield-Hartford
area and the Providence-Fall River area. No two state legislatures involved could alone establish a local government encompassing the interstate areas. Through compact two states could,
however, require the contiguous area governments of, e.g., Fall
River and Providence, to coordinate and cooperate in their
activities.

The dangers of nonresponsiveness of such governments may be
minimized by instituting the office of ombudsman into the SRA's.

2. Strengthening the New England Compact

At Stage Two, the constitutions of the New England states must be amended to permit at-large popular election of two state representatives to the Board of Regional Directors. The board, at this point, will be institutionalized as the New England Cabinet. A chief administrator will be elected from among the cabinet members for a two-year term. His successor must come from another state. This institutionalization will be necessary because of the growing importance of the former compact. It will also serve to democratize the compact in preparation for Stage Three.

At Stage Two, the Federal Constitution should be amended to permit state delegation of legislative authority to the regional government without retention by the states of legislative veto power. The caliber of the regional government's activities will not necessarily change; rather, decisions made at the regional level will become final.

Many devices used currently to implement governmental decisions will be used by the regional government. For instance, in the area of land-use planning, the tools of zoning, land acquisition, and taxation will be used. Other traditional methods of land-use control such as subdivision, regulation, business licensing, highway access, historic district regulation, etc., should assist in implementing regionwide plans. While such tools may ultimately prove inadequate, their use while new methods of implementation are developed will ease the transition.

3. Financing

At the core of successful planning is sufficient data on which to make reasonable decisions. Because New England's natural beauty is one of its greatest resources, land-use planning is crucial. Subregional governments at an early part of Stage Two will compile inventories of all land within their respective jurisdictions. Each existing tract of land will be classified and catalogued according to existing use, desired use, present owner, and full-value assessment. In the latter portions of

Stage Two, subregional land-use plans will be coordinated with
the prescribed regional land-use plan of the New England Cabinet.
The land-use plan will in turn be used as the basis for the
new tax to be described in Section V.

Upon establishment of the two metropolitan governments, the
payroll tax will be eliminated and an income tax initiated within the broader subregion. Progressive state income taxes, enacted after state constitutional amendment, should become the major
source of state revenue.[48] At this point, the local income tax
could be a fixed percentage of the state tax.[49] By "piggybacking"
the two taxes, the local tax should be easy to compute and administer. It would also avoid conflicts with the socially-
desired exemptions and credits set by the state.

Additional funding to both intra-and interstate organizations would come from the states and the federal government.
The federal government could impose a tax on the New England
Compact Region and return the funds to the states comprising
the region for allocation. This scheme avoids the legal pit-
falls present in state and federal constitutions when proposals
are made for state taxation at regional levels. The scheme
would be improved if the federal government distributed the tax
revenue directly to the regional and SRA levels. Ideally, it
is hoped that the New England region will ultimately assume its
own authority to tax and thus become self-sustaining.

Until the region does become self-sustaining, the traditional disposal of federal funds will continue greatly to influence the region's development. Among direct expenditures by
the federal government, which will have influence on the region,
are the location of federal installations, award of federal contracts, and subsidy of private industry. Indirect federal
spending will continue to enable state and local governments to
attempt fiscal projects otherwise impossible. One typical expenditure is that provided for under the Urban Mass Transportation Act[50] whereby cities may receive funds to develop an urban
mass transport system. Substantially identical programs exist

in the areas of urban renewal, pollution control, and regional
planning.[51]

To repeat, State Two, to be completed by the year 2000, has
as its specific proposals:

(1)   Institutionalization of the Metropolitan Government;

(2)   Establishment of SRA's;

(3)   Strengthening the Regional Compact;

(4)   City payroll tax becomes broader income tax;

(5)   State income tax made progressive and increased.

IV.   STAGE THREE

The third stage envisions the creation of a truly regional
New England government. It should be completed in the years
2015-2020, the years of the project termination.

Modifications of the magnitude proposed will invariably
meet strong resistance from currently-entrenched-interest groups.
However, we submit that the current ecological and social dis-
equilibrium warrants more than token patchwork responses from
our elected leaders. Only a callous and unconcerned official
could ignore the problems described in other chapters of this
report; and only an unenlightened and unresponsive official can
dismiss as academic rhetoric concerned efforts for meeting problems on the level--if not in the exact form--of our proposals.
We do not intend to overemphasize currently-popular cliches.
However, major institutional response must take place in a
planned progressive manner. Otherwise the current trend toward
federalization of problem-solving, with all of its attendant
inadequacies, and the emasculation of lower levels of government
will be accentuated. Thus even some of our most extreme proposals, such as constitutional amendments, must be viewed as viable
given the framework of a 50-year time span and the complexity of
the problems.

The major step to be taken at this stage is the structural

completion of the subregional and metropolitan governments which
cross state lines. Further, state and federal constitutional
amendments authorizing states to consolidate their powers into
regional governments would enable them to set up subregional
governments encompassing territory in two states. Therefore, a
Springfield-Hartford government, a Providence-Fall River government, and a new government encompassing parts of western Massachusetts and southern Vermont could be established under the New
England Cabinet.

The proposed federal constitutional amendment should state
in substance that:

> Nothing in this constitution shall prevent any given
> state or group of states from abdicating their sover-
> eignty to a binding regional form of government.
> Nothing in this amendment shall permit secession from
> the federal union.

Additional language would specify the manner whereby the
region would be congressionally represented at the federal level.

V.   PROPERTY TAX REFORM MEASURES

1. Introduction

We shall now shift this discussion to a more concrete example of the elements of the aforementioned institutional reform.
This example is the general property tax and, although intrinsically microcosmic, it is an essential element to any reform
package.

Governmental reform must include sound revenue-gathering
mechanisms. We support total family income as the appropriate
source of revenue for supporting regional government. However,
proposals substantially reforming existing governmental structures in New England must also take into account the currently
important revenue-gathering device--the general property tax.
Since its inception the general property tax has been maintained
despite its inability to conform to the generally-accepted
theories of equitable taxation, primarily because it has been
a very stable source of revenue. Nevertheless, today's general

economic climate prohibits effective implementation of the tax,
and we join with those tax theorists who are disenchanted with
the property tax as currently administered. However, we depart
from their camp insofar as complete abolition is advocated.
Its revenue function must be preserved insofar as the financing
of property-oriented services is concerned. Equally important,
we propose giving new emphasis to the policing and control mechanisms of the tax, thus utilizing it as an essential force
within a planned progression towards regional government. In
the property tax we find a powerful tool for retarding unre-
strained migration patterns; for policing effluent discharges;
and for establishing sound land-use incentives.

2. Discussion of the Concept

   Assume that Stage One of the governmental reform previously
proposed has been implemented, and that the New England state
legislatures have sanctioned a functioning interstate compact.
One of the initial tasks of the Board of Governors would be to
determine the most desirable and functional land-use allocation
within the region. Hopefully, this task will be accomplished
in an atmosphere (insofar as possible) free from local political
pressures. This area land-use determination of subregional
areas will be made irrespective of existing interstate political
borders. The factors to be considered include:

   (1) The natural geography;

   (2) Existing land use
        (a) Industrial
        (b) Population concentrations
        (c) Recreational and natural resources
        (d) Urban/rural interface.

   (3) Current trends and legislation concerning intrastate
        governmental consolidation efforts;

   (4) Existing interstate cooperation and consolidation
        efforts;

   (5) Existing federal land-use policies and restrictions;

(6)   Socially- and ecologically-desirable conditions
      (a)   Balanced urbanization
      (b)   Tolerable pollution levels
      (c)   Efficient regional transportation
      (d)   Accessibility of recreational and natural
            resources.

The general classifications of the subregional areas are
three in number, and might reflect percentage designations similar to those shown in Table 6.1 (it must be emphasized that

DESIGNATION	PERCENTAGE

Class One: URBAN

a. Public	30%
(1) Roads	
(2) Hospitals	
(3) Municipal and governmental buildings	
b. Industrial	30%
(1) Heavy (stressed)	
(2) Light	
(3) Service	
c. Housing	30%
d. Recreational and open	10%

Class Two: URBAN/RURAL

a. Public	20%
b. Industrial	25%
(1) Heavy	
(2) Light (stressed)	
(3) Service	
c. Housing	40%
d. Recreational and open	15%

Class Three: RURAL/NATURAL

a. Public	5%
b. Industrial	10%
c. Housing	30%
d. Recreational and open	55%

Table 6.1   Desired Land Use in the Year 2020--Subregional
            Area Designations and Percentages

the figures represented in this table have been arbitrarily selected for purposes of descriptive analysis).

For a state such as Massachusetts there would be approximately sixteen of these SRA's as shown in Figure 6.1. Of the sixteen, six would potentially transcend interstate borders as currently drawn. Three of these six are currently designated as

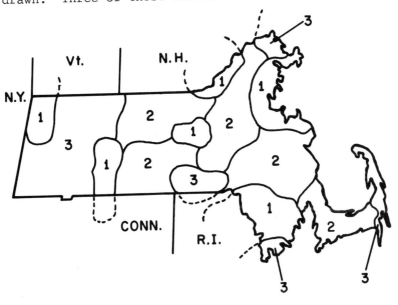

Figure 6.1 Massachusetts Sub-Regional Areas

interstate Standard Metropolitan Statistical Areas. The other
three would reflect a policy decision of the board relating to
desired land use in the region. Within any given SRA a state
would have autonomy to develop land as it chooses and where it
chooses, under the auspices of a local planning board and
subject to the absolute percentage requirements defined initially by the board.

As an illustration of the application of this land-use/
taxing concept, we might look at the hypothetical future development of the city of Bangor, Maine, situated as shown in

Figures 6.2 and 6.3. Assume that the SRA within which Bangor is situated has been designated to be an Urban/Rural-Class Two area, and that the current and desired land-use patterns in that SRA are characterized as shown in Table 6.2 and Figure 6.4.

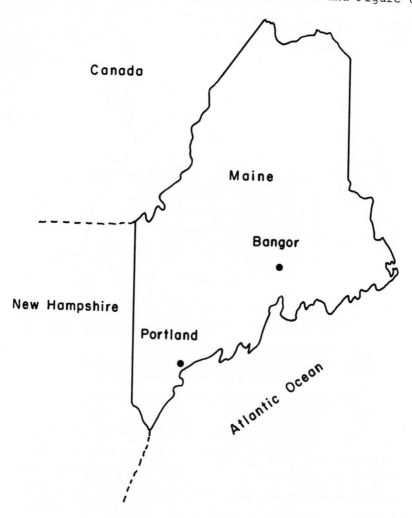

Figure 6.2 The State of Maine

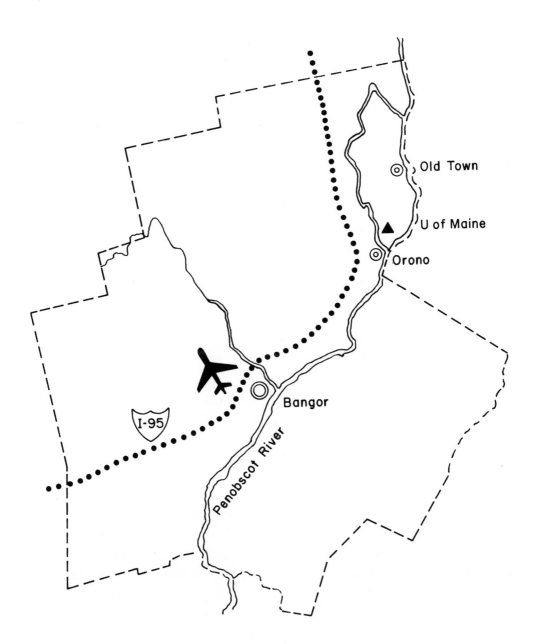

Figure 6.3   The Bangor Area

Land Designation	Existing Percentages	Desired Percentages	Percent Change
URBAN/RURAL: Class Two			
a. Public	10	20	plus 10
b. Industrial	10	25	plus 15
c. Housing	30	40	plus 10
d. Recreational	50	15	minus 35

Table 6.2   Subregional Area Example of Shifting Land-Use
Bangor, Maine

These exhibits indicate that, in order to actually effect
the change to the Urban/Rural classification, major incentives
and pressures will have to be applied. This can be accomplished
most efficiently and with a minimum of social and governmental
disruptions by the implementation of land-use controls such
as zoning and property taxes.

This technique would first require the regional board to
determine a primary tax rate structure for each SRA classification within a range of 0% to 25% to full-value assessment, as
shown in Table 6.3. No individual SRA would have the authority
to tax above these absolute amounts. (Again, the figures have
been arbitrarily determined for purposes of discussion.) Hopefully, the range within each category of land reflects feasible
limits on the regional level for influencing a) land-use decisions,
b) population concentrations, and c) industrial growth. In
addition, such designations might directly affect the standards
of quality of life by encouraging, for example, the heaviest
polluters to relocate to areas of less pollution where (presumably)
lower primary property tax rates would prevail.

Further, it should be noted that SRA's could be encouraged
to proceed more rapidly in the direction of the prescribed land-
use percentages by having the region tax publicly-held lands
as sort of a penalty for nonresponsiveness. For example, in the

LAND USE	URBAN (CLASS ONE)	URBAN/RURAL (CLASS TWO)	RURAL/NATURAL (CLASS THREE)	RANGE[4]
Industrial	20-25%	15-25%	10-25%	25%
Housing[2]	15-20%	10-20%	5-20%	↑
Recreational and open (private)	10-15%	5-15%	0-15%	
Recreational and open[3] (public)	0-5%	0-5%	0-5%	↓ 0%

1. To be determined by the Regional Board of Governors.
2. One-half acre per family to be permitted as a maximum at these basic rates.
3. Tax upon some publicly-held land might be deemed advisable in some instances where quasi-penalties are deemed necessary.
4. The tax rate will never be permitted to exceed 25% of full assessed value, as control above this level will be accomplished through zoning procedures.

Table 6.3 Absolute Tax Rate Variables[1] in Percentages of Fair Market Value and at Full Assessment

Bangor area the hypothetical figures show that an eventual decrease of 35% in recreational and open lands has been deemed acceptable and desirable within the subregion. By taxing public surplus holdings of land, pressure could be applied to the local implementing governments to reallocate public land, for example, towards housing. This reallocation could be accomplished by sale to private developers and financed through the regional development corporations, as previously discussed. The region's absolute rates place the highest premium on industrial land in an Urban Class One Subregion. Disregarding public holdings of land (which may or may not be subject to a tax), the lowest rate is applied to private undeveloped land in a Rural/Natural area. Consequently, this device makes it necessary for taxpayers to value carefully the advantages of a specific location in relation to personal finances or corporate profits. These rates reflect the highest level at which government tax policy can determine land-use within a given area. Any further land-use control efforts should stem from the standard zoning procedures.

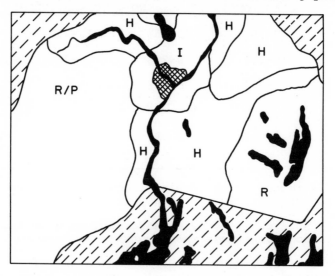

Key:

▨	Downtown Area
▢	Class Two: Urban/Rural
▨	Class Three: Rural/Natural

P - Public   I - Industrial
R - Recreational   H - Housing

Figure 6.4 Land Use Designations in the Bangor Area

Equal revenue needs are neither anticipated nor desired
for each class of SRA, for an individual subregion should be
able to provide the services that its citizens may desire.
Therefore, it should be obvious that some mechanism must be
employed whereby an SRA can determine, within the absolute limits
just discussed, what its specific rate will be. We submit that
such a control can best be exercised by creating a mixture of
tax credits and deductions that would tend to encourage the

growth desired within each SRA, and finance the services.

This secondary level of tax control, as shown in Table 6.4, for the Bangor region, is to be imposed at the subregional level and involves a sophisticated manipulation of a system of credits and deductions. This system should be designed not only to control subregional land use, but also partially to discourage

Land-Use Classifications	Deductions	Credits	Class Two Absolute Limits
Heavy Polluter Moderate Polluter Light Polluter Light Industry Banking and Invest. Service Small Service Commercial Farm	Flexible Variables		Industry/Business 15-25%
Multi-Family (>10) Multi-Family (3-10) Single (>3 C's) Single (w/o C's) Single (3 C's) Single (2 C's)			Housing 10-20%
Urban Area (+ 1/2 acre) Rural Area (1/2 acre) Wild (+ 15 acres)			Private Undeveloped 5-15%
Public Undeveloped Public Developed	Taxed at Regional Level		Public 0-15%

Table 6.4 Secondary Tax Control at Subregional Level-- Bangor

overpopulation and pollution, currently two of our most acute social evils. Indirectly, this system should also encourage the redistribution of single-family housing units to those couples currently having children living within the family social unit, who presumably need more house and yard space than do older

or single persons. Of course, an absolute desire to maintain
these privileges or luxuries can be indulged by paying a tax
premium. Such incentives can also help to relieve some of the
current pressures on public education by allowing certain communities within a subregion to concentrate upon providing school
and teaching facilities. In all instances, the credit/deduction
scheme as well as absolute rates should be subject to periodic
review in order to insure flexibility and response to changing
conditions as reflected by increased demands for government
services.

3. Conclusion

In concluding this discussion of our novel tax system, it
should be emphasized that, although our plans and proposals are
untested, the mechanisms for implementation are available to
governments under our presently-constituted institutional system. As governments eventually consolidate, this tax scheme is
potentially of great utility in instituting a land-use policy
which will conform with, and be fiscally responsive to, the
future needs of our society.

REFERENCES

1. Perloff, Dunn, Lampard, and Muth, "Regions, Resources and
    Economic Growth," 19 (1960). The regional concept of
    states within the United States seems never to have been
    formally agreed upon or officially designated. These
    authors have divided the country into eight regions for
    purposes of statistical consideration. These are: New
    England, Middle Atlantic, Great Lakes, Southeast, Plains,
    Southwest, Mountain, and Far West. The classification is
    somewhat arbitrarily geographic, and should not be taken
    as a firm diecast for the proposed regional governmental
    structure suggested infra.

2. Advisory Commission on Intergovernmental Relations, Urban
    and Rural America: Policies for Future Growth, Appendix I-B
    at 175 (1968).

3. Advisory Commission on Intergovernmental Relations, Fiscal
    Balance in the American Federal System, Vol. I, Report
    No. A-31, at 60.

4. Advisory Commission on Intergovernmental Relations, Urban
    America and the Federal System, Report M-47, at 1, October
    1969.

5. Advisory Commission on Intergovernmental Relations, State
    and Local Finances: Significant Features 1967-70, Report
    No. M-50, Tables 14-17. [Hereinafter referred to as
    ACIR M-50.]

6. ACIR A-31(I), supra note 3, pages 90 et seq. and Table 12.

7. Part II, ch. 1, s/s 1, art. 4.

8. Massachusetts General Laws, ch. 59, s/s 38.

9. Boston Gas Co. v. Assessors of Boston, 334 Mass. 546, 566.

10. Bennett v. Board of Assessors of Whitman, 354 Mass. 239
    (1968).

11. J. Shannon, "Conflict Between State Assessment Laws and
    Local Assessment Practice," in Property Taxation U.S.A.
    (1967) 41. [Hereinafter cited as "Shannon."]

12. Advisory Commission on Intergovernmental Relations, Fiscal
    Balance in the American Federal System, Vol. II, Report
    No. A-31 at 123. [Hereinafter referred to as ACIR A-31(II).]

13. Ibid. at 102.

14. Ibid. at 103.

15. Ibid. at 123-24.

16. Winters, "Interstate Metropolitan Areas 5," Michigan Legal
     Publications (1962).

17. ACIR, supra note 5, at 2.

18. Christensen, Land Use Control for the New Community,
     6 Harv. J. Legis. 496, 506 (1969).

19. See, e.g., New Hampshire Rev. Stat. Ann., chs. 12 and 36
     (1966); Rhode Island Gen. Laws 45-22.1 (1968); Vermont Stat.
     Ann., ch. 10, sec. 301 (1958).

20. Dixon, "Constitutional Basis for Regionalism: Centrali-
     zation; Interstate Compacts; Federal Regional Taxation,"
     33 Geo. Wash. Law Review, 47 (1964).

21. Ibid. at 54, 55.

22. See S. Scott and J. Bollens, Government: Regional Organiza-
     tion for Bay Conservation and Development (1967).

23. See Koerner, Interlocal Cooperation: The Missouri Approach,
     33 Missouri Law Review 442 (1968).

24. Advisory Committee on Intergovernmental Relations, A Handbook
     for Interlocal Agreements and Contracts (1967).

25. See, e.g., statutory authorization for interlocal coopera-
     tion commissions, Rhode Island Gen. Laws, sec. 45-40-1 (1968).

26. For a general discussion of the legal classification of
     special districts, see Novak, Legal Classification of Special
     District Corporate Form in Colorado, 45 Denver L.J. 347.

27. Massachusetts Constitution Amended, Art. II, sec. 8.

28. For an earlier plan for consolidation of Boston Metro-
     politan Area Governments, see Planning for the Greater
     Boston Metropolitan Area, 5 Pub. Ad. Rev. 113 (1945).

29. Dixon, supra note 20, at 57.

30. U.S. Const., Art. I, sec. 10. See also, Frankfurter
     and Landis, The Compact Clause of the Constitution - A
     Study in Interstate Adjustments, 34 Yale L.J. 691 n. 25
     (1925). "The Constitution puts this power negatively
     in order to express the limitation upon its exercise
     by putting this authority for State action in a section
     dealing with restrictions upon the States. The signi-
     ficance of what was granted has probably been consider-
     ably minimized."

31. Dixon, *supra* note 20, at 57 *et seq.*

32. ACIR, *supra* note 4, at 82.

33. The utility of such federal spending as a means for
     promoting regional development has been recognized by
     both former Secretary of Agriculture Orville L. Freeman,
     "Towards a National Policy on Balanced Communities,"
     53 Minn. L. Rev. 1163, 1174 (1969) and John H. Southern,
     in a speech on Regional Growth and Development, before
     the 42d Annual Agricultural Outlook Conference, Nov. 19,
     1964.

34. See Staff of Senate Comm. on Banking and Currency, 85th
     Cong., 2d Sess., "Development Corporations and Authori-
     ties," vii. (Comm. Print 1958); U.S. Small Business Admin-
     istration, "Development Credit Corporations" (1954).

35. Note, "Legal Limitations on Public Inducements to Indus-
     trial Location," 59 Columbia L. Rev. 618, at 639-40. See
     generally H.R. Rep. No. 1889, 85th Cong., 2d Sess. 69
     (1958); U.S. Small Business Administration, *supra*
     note 34, at 4-6.

36. Massachusetts Business Development Corporation.

37. Mass. Acts and Resolves 1953, ch. 671, sec. 4.

38. See *Opinion of the Justices*, 211 Mass. 624, 98
     N.E. 611 (1912); *Opinion of the Justices*, 204 Mass.
     607, 91 N.E. 405 (1910). See also Mass. Const. Declara-
     tion of Rights, art. X, Mass. Const., pt. II, art. IV,
     ch. 1, '1; Mass. Const., art. LXII (Commonwealth credit
     not to be pledged for the benefit of private enterprise).

39. Note *supra* note 35, at 640. The constitutional objec-
     tions relate to the expenditure of public monies for
     private benefit. See generally Tilden, *supra* note 33,
     at 8-9. Cf. *Ayer v. Commissioner of Administration*,
     340 Mass. 586, 165 N.E. 2d 885 (1960) which ruled that
     a "membership corporation" was solely functioning to
     serve the Commonwealth and therefore subject to consti-
     tutional limitations on Commonwealth action.

40. ACIR A-31(I), *supra* note 5, at 108.

41. *Ibid.*

42. Elisin, "The Fineness of Metropolitan Areas," Michigan L.
     School Legis. Res. Ctr., 32 (1964).

43. See Blum and Kalsen, "The Uneasy Case for Progressive
     Taxation," 19 U. Chicago L. Rev. 417, 520 (1952).

44. CCH State Tax Guide at 1533.

45. Walker, Fiscal Aspects of Metropolitan Regional Devel-
     opment, 105 U. Pennsylvania L. Rev. 489, 497 (1957); see also
     Hartman, Municipal Income Taxation, 31 Rocky Mountain
     L. Rev. 123 (1958).

46. Massachusetts Constitution Amended, Art. 44, Opinion
     of the Justices, 266 Mass. 583, 586.

47. This may be required by Avery v. Midland County, 390
     U.S. 474 (1968).

48. See W. Newhouse, Constitutional Uniformity and Equality
     in State Taxation, 178-80 (1959).

49. This is the system currently used by Vermont which
     takes a flat percentage of the federal tax rate paid.
     Vermont Statutes Annotated Secs. 32:5822.

50. 49 U.S.C., sec. 1601-30 (1964), as amended by 49 U.S.C.A.,
     secs. 1602-07, 1610, 1611 (1969). See generally Fitch,
     Urban Transportation and Public Policy, 227 (1964).

51. See, e.g., Federal-Aid Highway Act, 23 U.S.C., sec. 134
     (1964), requiring the existence of a regional highway
     plan as a prerequisite to financial assistance.

INDEX